S0-AUG-155

OPTIMAL EXPANSION OF A WATER RESOURCES SYSTEM

WATER POLLUTION

A Series of Monographs

D. T. O'Laoghaire and D. M. Himmelblau. *Optimal Expansion of a Water Resources System*. 1974

IN PREPARATION

C. W. Hart and Samuel L. H. Fuller (eds.). *Pollution Ecology of Freshwater Invertebrates*

OPTIMAL EXPANSION OF A WATER RESOURCES SYSTEM

D. T. O'LAOGHAIRE

Departamento de Matemáticas
y Ciencia de la Computación
Universidad Simon Bolivar
Caracas, Venezuela

D. M. HIMMELBLAU

Department of Chemical Engineering
The University of Texas
Austin, Texas

ACADEMIC PRESS New York and London 1974

A Subsidiary of Harcourt Brace Jovanovich, Publishers

To

Seán *Betty*

and

Siobhán

ACADEMIC PRESS, INC.
111 Fifth Avenue, New York, New York 10003

United Kingdom Edition published by
ACADEMIC PRESS, INC. (LONDON) LTD.
24/28 Oval Road, London NW1

Library of Congress Cataloging in Publication Data

O'Laoghaire, D T
Optimal expansion of a water resources system.

Includes bibliographies.
1. Water resources development–Mathematical models. 2. Mathematical optimization. I. Himmelblau, D. M., joint author. II. Title.
TC409.04 628.1'1 73-7440
ISBN 0–12–525450–4

PRINTED IN THE UNITED STATES OF AMERICA

CONTENTS

Chapter 4 Application of the Optimization Algorithm to a Water Resources System

Chapter 5 The Sensitivity of Planning Decisions in River Basin Management

Chapter 6 How to Incorporate Water Quality and Pollution Considerations into the Model of the Water Resources System

Appendix A Computer Program Listing and Instructions

FOREWORD

This series deals with methods of ensuring supplies of pure water to areas which have polluted water. Criteria for purity and pollution vary with the intended use of the water, but there is no question that in the recent past water pollution has become worse while, on the other hand, purity requirements for several important uses have become more stringent. Hence the need for action on a national level has been recognized in many countries.

In the United States, the 1972 Amendments to the Federal Water Pollution Control Act made water pollution control an important and permanent governmental requirement. Waste water discharged into a city sewer or into a receiving water body must be adequately treated. The requirements are specified in detail in the Act; industries and municipalities are now in the process of taking corrective actions which are often quite expensive. Similar laws are either in preparation or already on the books in other countries.

The series as a whole addresses itself to water supply planners who wish to have on their shelves a comprehensive and practical multivolume manual on water pollution problems. While the solution of any water pollution problem must in general be tailored to the specific situation in which the problem arises, much can be learned from the experience of others who have already solved similar problems, or at least developed technical methods for solutions. A number of good books on water pollution subjects have been available in the past, but now that the technology and legislation are beginning to change rapidly, it is necessary that the workers in this field have at their disposal a series of volumes covering the most current state of knowledge. This series deals primarily with engineering solutions, but since technical feasibility alone does not guarantee adoption of any scheme, volumes dealing with selected other aspects of the wide spectrum of water management are also planned.

The series is open-ended because the development of ever more sophisticated industries and life patterns continuously creates new water pollution problems and, hopefully, their solutions. By dealing with water *quality* it is designed to encourage its users to contribute to the quality of life in general.

This first volume by Professors Himmelblau and O'Laoghaire is particularly suitable to open the series, because it deals with planning for maximum multiuse of water resources. We welcome the reader to this series and hope that he will find the most current information in this volume and in the volumes to come.

K. S. Spiegler
J. I. Bregman

PREFACE

How to carry out an optimal expansion of an existing water resources system is of continuing importance because of the rising demand for and limited supply of water in many areas of the world, particularly in the southwestern part of the United States of America. Governmental agencies in the U.S.A. and elsewhere have made large public investments in the field of water resources in the past and will continue to do so in the future. Whenever investment in a water resource project is under consideration, important questions such as what is the economic value of the project(s), what is the optimal scale of development of the project(s), and when should the project(s) be constructed need to be answered. It is only through the use of an analytical economic evaluation that competitive uses for capital can be quantitatively evaluated.

This monograph describes a methodology that can be used in water resources planning taking into account both water quantity and quality while still remaining computationally tractable. It is concerned with the optimal expansion of a realistic water resources system to meet an increasing demand for municipal and industrial use, irrigation, energy, and recreation over a planning horizon of T_{max} years. Although the problem of quantitatively describing a water resources system in a realistic fashion is forbidding, the outlook for quantitative analysis is good. Some of the problems of describing a river basin include:

1. We have only a fragmentary knowledge of the relevant parameters to include in a river basin model.
2. We do not know how important some of the variables are in relation to others.

3. We do not know which are the most significant parameters in any model in influencing the model outputs.

Nevertheless, by formulating models of river basins that mesh successfully with the available optimization techniques, and by analyzing and improving the models, these difficulties can be ameliorated.

In formulating the model of the river basin it is assumed that a number of possible dam sites are available for the further regulation of imported waters into the basin. The model has been limited to systems that have (1) deterministic inputs, (2) a network configuration, (3) linear constraints, and (4) capital investment and operating decisions made on a yearly and a monthly basis, respectively, so that the operating policy and construction policy could be optimized. The model of the system did not include (1) stochastic effects or (2) intangible benefits and costs that could not be quantified. Emphasis in the preparation of the model has been placed on the diversity of applicability rather than a specific river basin.

After discussion of the criteria for and scope of the problem of expanding an existing water resources system in Chapter 1, a water resources system model is developed and explained in Chapter 2. In Chapter 3 an optimization strategy is developed to maximize, over the set of alternative projects, the sum of the discounted present value of net earnings of the system subject to the water demands and various institutional, physical, and budgetary limits. The optimization problem is posed as a 0-1 mixed integer programming problem that is decomposed into the set of all feasible combinations, a capital budgeting problem; and the economic return is determined for each combination, an operating policy problem.

In Chapter 4 an example problem is formulated, solved, and discussed. The efficacy of the optimization algorithm is demonstrated by applying it to the solution of a capital investment problem in a model river basin that resembles a real river basin (the Maule River Basin in central Chile).

Chapter 5 briefly shows how to carry out a sensitivity analysis on a water resources system to discover the critical parameters and inputs in the model, parameters whose values in principle have to be obtained with the greatest accuracy. Chapter 6 indicates how water quality can be incorporated into the water quantity model. A FORTRAN listing of the computer program to execute the optimization algorithm will be found in Appendix A. For those not familiar with or who desire more information about existing techniques in optimization, we provide references at appropriate places in this text to the introductory book by Beveridge and Schechter, "Optimization Theory and Practice," McGraw-Hill, New York, 1970.

The approach and methodology developed in this monograph are intended to provide guidance to policy and decision makers. It is intended to isolate the economic effects of interrelated factors of water quantity and quality more explicitly to the end that water resources planning may more effectively and efficiently serve the needs of society. Probably one of the major flaws in the systems approach to water resources development is the inability of the scientist and engineer to provide the political decision makers with meaningful plans. It is hoped that this monograph will assist them in their task in the future.

The authors gratefully acknowledge the assistance of Professor William Lesso and Mr. James Lindsay in connection with the development of the computer programs used in this monograph. The work upon which this publication is based was supported by funds provided by the United States Department of the Interior as authorized under the Water Resources Research Act of 1964 as amended. D. T. O'Laoghaire would also like to express his appreciation to The Institute for Industrial Research and Standards, Dublin, Ireland, for financial support and to Dean Hermes Espinoza Sosa of the Universidad de los Andes, Mérida, Venezuela, for arranging for him to have sufficient time for writing the book.

Chapter 1

INTRODUCTION

During the last few decades the planning for the development of water resources has progressed from single-purpose projects, such as for irrigation or flood control, to multipurpose programs encompassing entire river basins. The rapidly increasing population of the United States and other countries of the world, together with the resource-consuming characteristics of our present society, makes it imperative that we improve the quality of our decision making for allocating public and private investments in the water resources area. Of particular importance is the precedence ordering and sizing of the construction of units in a water resources system over time. If a unit is constructed too soon, the cost of interest on the unused investment for even a few years can be substantial. On the other hand, failure to meet expected deliveries for firm hydroelectric energy and water as specified in contracts can result in even greater losses of system revenues and project benefits. This chapter discusses what the problems in planning for a comprehensive water resources system are, how they are attacked, and what criteria are used to evaluate the expansion of a water resources system.

1.1. Development of Water Resources

Comprehensive plans for the construction and operation of dams, canals, and so on have been completed or are being implemented in a number of countries, including Austria, Bulgaria, Great Britain, Israel, Japan, the Netherlands, South Africa, West Germany, and the United States. These plans do not necessarily cover all aspects of water utilization; some aspects are and will remain of no interest to certain countries, and hence one or more phases of planning cannot help but be subordinated to others that are more vital. Thus, plans in Japan are designated as comprehensive even if they relate only to river flow control, irrigation, drinking and industrial water, and hydroelectric energy. Planning for the entire field of water resources is most evident in those countries where dense populations and advanced industrial development prevent a balance of water use from being maintained now or in the near future in many of their river basins or regions.

If we look at a highly developed industrial country such as the United States, we find a correspondingly sophisticated development of its water resources. Water is used for (1) irrigation projects to provide food for an expanding population and reduce the need for imports, (2) navigation projects to improve the transportation network, (3) public-health projects to reduce or eliminate the incidence of water-borne diseases and improve the physical well-being of the population, and (4) hydroelectric energy projects to provide a cheap, renewable source of energy for an expanding industrial sector. In the arid western states the need for water for agricultural purposes has always exceeded the rainfall supply; the rapid depletion of ground-water supplies has animated the debate about the "shortage" of present water resources in these areas to meet future agricultural demands. Rather complex schemes have been proposed to transfer excess water to the areas of water shortage.

California, for example, is building an aqueduct to transfer water from the northern part of the state to the water-short southern part [California State Water Resources Board, 1957].‡ Texas has proposed the construction of a huge network of canals and reservoirs to transfer approximately 12 million acre-ft from East Texas and the Mississippi River Basin to the High Plains of West Texas and Eastern New Mexico [Texas Water Development Board, 1968] as shown in Figs. 1.1 and 1.2. Furthermore, a proposal has been made to divert and transfer waters that now flow into the northern Pacific Ocean, the Arctic Ocean, and Hudson Bay to the water-short areas of Canada, the United States, and Mexico [Parsons, 1964].

Projection of future water demands in the United States can be broken down for convenience into urban, industrial, and agricultural demands. Howe [1971], in predicting urban water demands, examined residential, commercial, industrial, and public uses of water, technological develop-

‡ References will be found at the end of the book. Supplementary readings, divided by chapter, are also listed there.

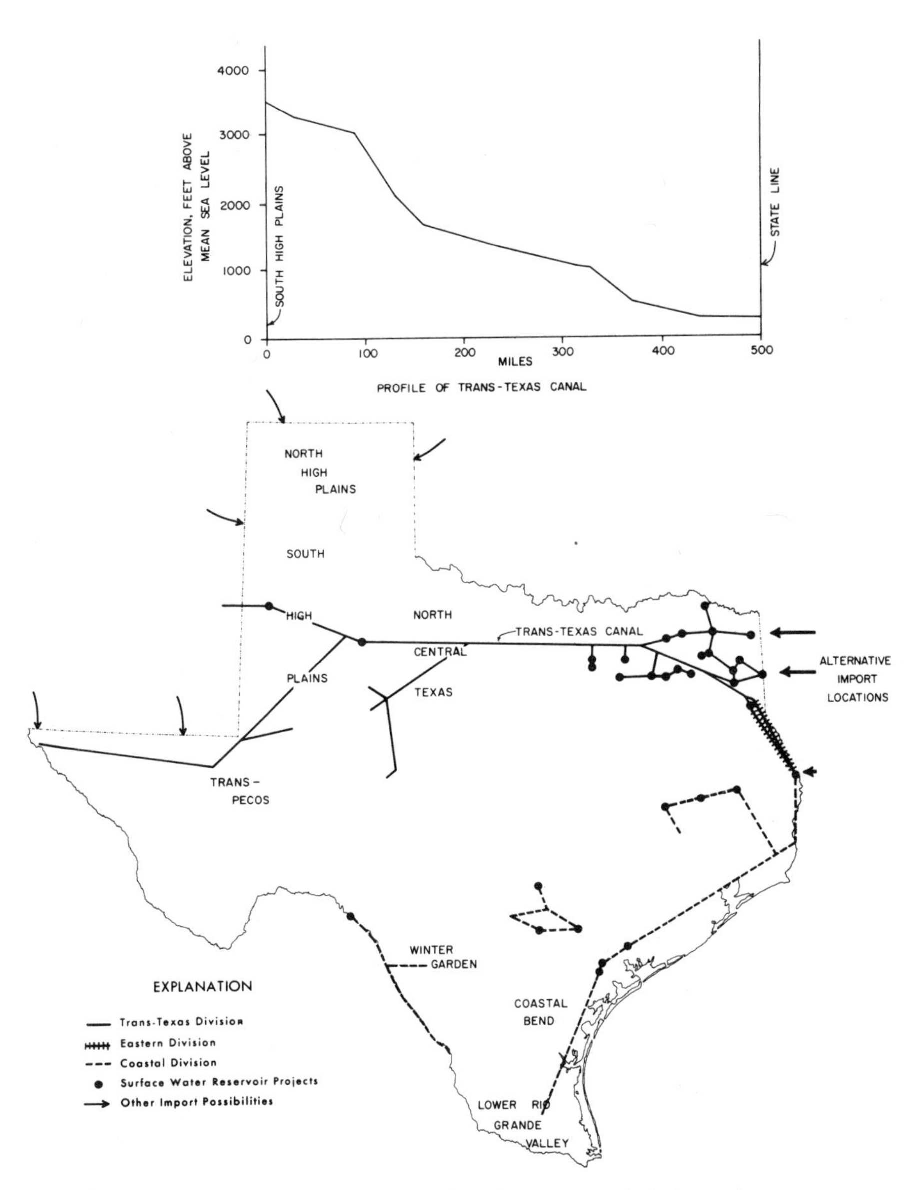

Fig. 1.1 Schematic diagram of the Texas water system (includes major conveyance facilities and related reservoirs). (From Texas Water Development Board [1971].)

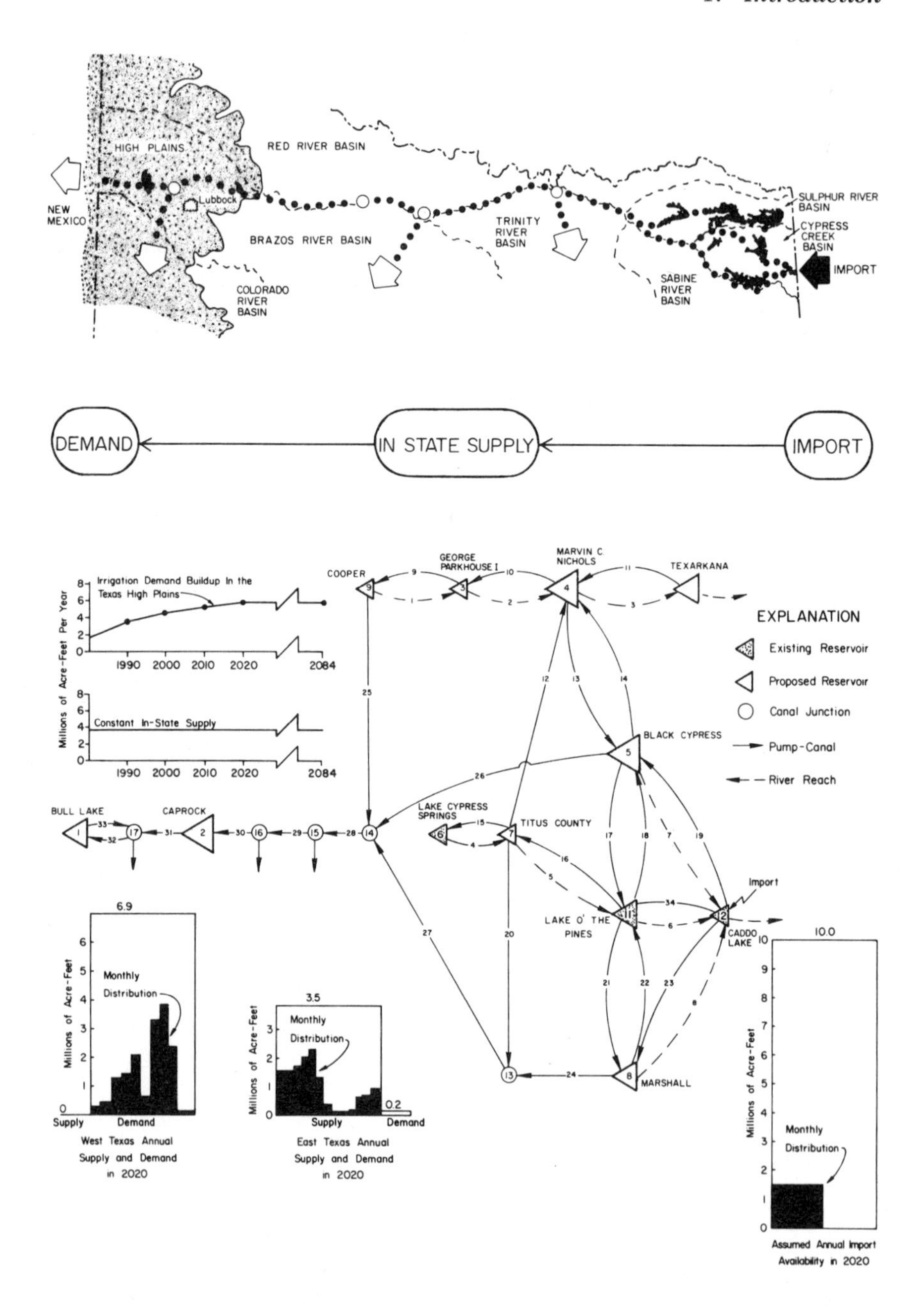

Fig. 1.2 Network flow diagram of the Texas water system and projected supply and demand relationships. (From Texas Water Development Board [1971].)

ments and costs of water-saving devices, additions to the housing stock by various categories, municipal water system loss and repairs, and other pertinent factors.

Future national residential water demands, currently the largest fraction of the total urban water usage (30–60%, depending on location), depend on price and are expected to range from an *additional* 4.6×10^9 gal/yr (14×10^6 acre-ft) at $0.20/gal to 2.7×10^9 gal/yr (8.3×10^6 acre-ft) at $1.20/gal. Of this about 30% will be consumptive use and 70% returned via the sewage systems. The Water Resources Council [1969] estimates that the increase by 1990 for municipal and rural domestic usage will be 5.9×10^9 gal/yr. To put the forecasts in the proper perspective, an increase of 4.0×10^9 gal/yr represents a 70–80% increase over the 1965 level of residential usage and a 47% increase in the total municipal production.

Commercial uses (hotels, restaurants, car washes, laundries, hospitals, golf courses, and so on) represent 10–25% of the urban water demands and are expected to be roughly the same in 1990.

Industrial demands (7–40% of urban water usage, depending on location) for 1990 are much more difficult to predict because they depend on future cost of water, effluent standards, costs of improving water quality, and type of heat discharge. Howe forecast that withdrawals of water for thermal electric generation would be 7.6×10^{13} gal/yr in 1990 (versus 5.0×10^{13} in 1970) and that consumptive use in 1990 would be approximately 2.0×10^{12} gal/yr (versus 1.4×10^{11} in 1970). The figures for consumptive use differ somewhat from those of the Water Resources Council (1968) of 4.3×10^{11} (1970) and 1.3×10^{12} (1990). Indications for some other industries are that more stringent pollution controls will prevent water usage from increasing substantially.

As to agricultural demands, public policies in 1990 with regard to support prices for agricultural commodities and low-cost irrigation water may change with increasing urbanization of the United States population. Water uses for agriculture are quite sensitive to these policies as well as to farm and processing technology. Higher prices for irrigation water in the arid regions of the West would drastically reduce the agricultural usage of water in 1990. For example, Howe estimated that the water usage of a typical 860 acre farm would be reduced from 5600 acre-ft/yr to 1100 acre-ft/yr, a reduction of 80%, if the cost of water went from the current $2–3 to $25/acre-ft. Since public investments for water in Southern California run close to $100/acre-ft, there may be some tendency to increase irrigation water costs in the future. Reduction in price supports for wheat, feed grains, and cotton would have relatively little effect on irrigation water demands. Because irrigation of crops currently accounts for over

80% of the consumptive use of water in the United States, mostly in the West, the development of new water facilities will be markedly influenced by farm product prices and by future politics and is not easily predicted.

In less industrialized nations there is a similar interest in the development of water resources. In Colombia, for example, only 1.18% of potential hydroelectric resources had been developed in 1950 [IBRD, 1952]. By 1968, however, the installed generating capacity had been increased eightfold [Colombia Information Service, 1969]. Projects of the size and sophistication of the California Water Plan are under way in the Departamento del Chocó and have been proposed for the Caquetá River Basin (part of the Amazon Basin).

Some countries have less need to develop their water resources than others. Although the supply of water in a given region may be adequate now, however, the supply varies in time and place only within small limits depending on the hydrological and meteorological conditions of the region. The demand for water, on the other hand, is determined by population, social customs, and the need to develop food and industrial production in line with present and future population growth and economic requirements. The demand for water is everywhere on the increase and is thus becoming increasingly difficult to satisfy. In northern countries bordering on the polar region—with their high precipitation, numerous lakes and rivers, and low density of population—there is no cause yet for concern, except perhaps in a few centers such as large cities. Other countries, however, currently find themselves in dire straits with regard to the supply of water, and naturally tend to have sophisticated planning and efficient organization with regard to future development. South Africa, Israel, and Spain are typical examples.

Enormous capital investments are being made and contemplated for the future for water resources projects of increasing size. Through 1966 the International Bank for Reconstruction and Development (IBRD) gave $6 billion for projects that serve to provide adequate supplies and management of fresh water [IBRD, 1967]. In the same period the Inter-American Development Bank [Carter *et al.*, 1967] lent one third of its resources for water development projects. This trend will be continued in the next decade, when $30.5 billion will be spent on worldwide dam construction projects alone [McQuade, 1970].

Table 1.1 summarizes, by time periods, the additional facilities and estimated initial facility construction costs for a typical large river system, the Ohio. Category A in the table shows the primary areas of water resource development for stream flow control and in-place use for flood control, water supply, stream quality control, navigation, and hydroelectric

Table 1.1

Forecast of Capital Expenditures for the Ohio River Basin Development Program[a]

	Cumulative, in addition to 1965 program			
	To 1980		To 2020	
	Amount	Cost[b] (billion dollars)	Amount	Cost[b] (billion dollars)
A. Water resource programs (stream flow control and in-place use)				
1. Flood Control				
a. Storage (million acre-feet)	10.0	2.51	33.4	8.50
b. Local protection projects (miles)	152	0.15	488	0.34
c. Channel improvement (miles)	2,394	0.09	6,328	0.24
d. Flood plain information studies (number of studies)	200	0.01	700	0.02
2. Low flow control (million acre-feet)	5.8	1.40	16.1	3.96
3. Navigation				
a. Improvement to existing system (miles of channel)	2,187	0.46	2,187	0.67
b. Potential extensions and new waterways (miles of channel)	172	0.01	527	1.14
4. Hydroelectric power (megawatts)	7,200	0.81	40,000	4.50
Total, water resource programs cost		5.44		19.37
B. Related programs				
1. Outdoor recreation, sport fishing, and hunting (million man-days)	132.7	0.46	494.6	1.72
2. Watershed land treatment and management (million acres)	11.4	0.29	29.2	0.73
3. Lands to be irrigated (million acres)	0.1	0.01	1.3	0.12
4. Lands to be drained (million acres)	3.2	0.42	4.0	0.54
Total, related programs cost		1.18		3.11
Grand total, water and related land framework program cost		6.62		22.48

[a] Source: Drobny [1970].
[b] In 1965 dollars.

energy. Structural features of upstream watershed projects, lands for recreation, and all other appurtenances to water resource development projects are included in the cost estimates.

To optimize the benefits from the invested capital, projects cannot be evaluated by considering each element in the project individually. Projects can be envisioned, planned for, and designed correctly only when they are considered as components of the overall development within a region. In the context of the development of water resources, unified development is achieved by multiunit, multipurpose systems; that is, each unit of the system, such as a dam, can serve many purposes—irrigation, energy, flood control, and recreation—and also can be combined with other units in the system to serve the total demands on the system.

For a plan to be really comprehensive, it should aim at the optimum development of all resources of a river basin, including land, water, and other natural resources. The idea that the unified development of the water and other natural resources of a drainage basin could promote economic development and positive social change was first demonstrated in the Tennessee Valley under the aegis of the Tennessee Valley Authority (TVA) [White, 1965]. The TVA planned and executed the first large-scale multiunit, multipurpose river development in the world [Maass, 1962]. Flood control, navigation, and hydroelectric energy generation were the manifest purposes of the development. The success of the TVA's work has led to the foundation of similar authorities in countries as disparate as India, Pakistan, Colombia, and Mexico [Barkin and King, 1970; Posada and de Posada, 1966; Sen, 1969].

The unified development of water resources has gained general acceptance at the national and state levels in the United States [U.S. Congress Senate, 1962; California State Water Resources Board, 1957; Texas Water Development Board, 1968]. This concept has been defined by the President's Water Resources Council:

> River basins are usually the most appropriate geographical units for planning the use and development of water and related land resources in a way that will realize fully the advantage of multiple use, reconcile competitive uses, and coordinate mutual responsibilities of different agencies and levels of government and other interests concerned with water use. . . . Despite this primary confinement to an area, the fact should be recognized that planning also requires consideration of pertinent physical, economic, and social factors beyond the area.
>
> To the extent feasible, programs and projects shall be formulated as part of a comprehensive plan for a river basin or other area, and the report proposing development shall indicate the relationship to the comprehensive plan. When a program or project has been formulated independently and not as part of a comprehensive plan, the report shall indicate, to the extent practicable, the relationship

of the program or project to the probable later developments needed or to be undertaken in the basin and the reasons for proposing to proceed with the proposed program or project independently. [U.S. Congress, Senate, 1962]

One of the primary reasons why governments at all levels, rather than private companies, have accepted responsibility for water resources development is that water resources projects are capital-intensive and have long gestation periods, making them unattractive for private investments. Perhaps a more important reason is that water resources projects have very large indirect or secondary benefits that cannot be captured by private investors. Thus most governmental agencies as well as the public have accepted the hypothesis that water resources development leads to a great many supplementary benefits and opportunities.

It should be recognized that promotion of economic development through water resources projects may have caused an overinvestment in these fields in many countries [Tinbergen, 1964]. As explained by Hirshleifer *et al.* [1960], the reasons for the prevalence of overinvestment in water resources are complex, but certainly other factors besides economic development come into play. One possible explanation is that the policy maker(s) have personal preferences for monumental projects, so that they will be heroes not only in their own time but to later generations.

1.2. The Systems Approach to Solving Water Resources Problems

In recent years governmental agencies as well as private organizations have sought to cope more successfully with large, complex problems in the area of water resources by systems analysis. A *system* in general is an arbitrarily isolated combination of elements (abstract and arbitrary subdivisions) of the real world. Usually the elements correspond to physical components of the real world, as illustrated in Fig. 1.3 for a river basin—components such as rivers, dams, sources of water, and users of water. The mathematical representation of the system is termed the (mathematical) *model* of the system. It would be misleading to offer a definition of the systems approach with the pretense that it would be universally accepted. Nevertheless, it is generally agreed that the systems approach represents an attempt to find answers to questions that are posed regarding complex assemblies of physical systems with interaction between the subsystems. Systems analysis is undertaken in order to make rational decisions insofar as possible as to the optimal design, selection, or operation of a physical system. As might be expected, systems analysis is primarily useful

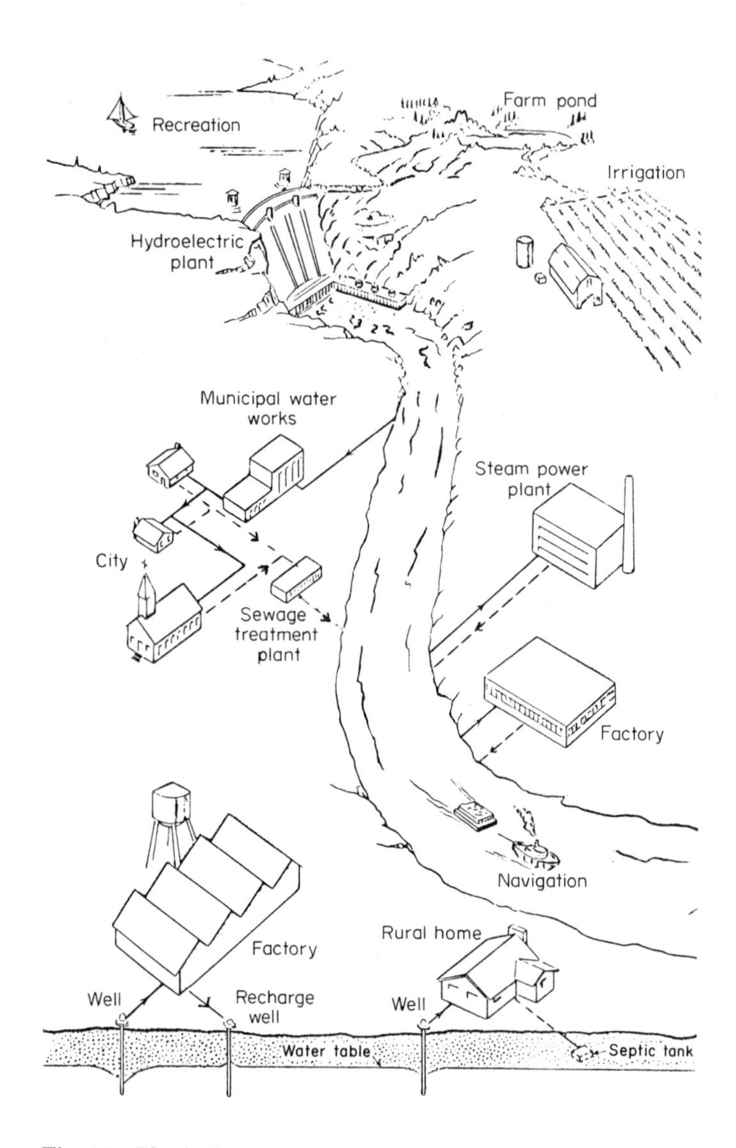

Fig. 1.3 Physical components of a river basin that form a system.

in dealing with planning problems that are sufficiently complex that no one man can be considered an "expert" on every aspect of the situation. In dealing with such problems it should be emphasized that systems analysis cannot replace experts in the appropriate disciplines any more than they can replace the policy makers or decision makers themselves.

To provide a better perspective on the systems approach for a water resources system, we can inquire as to how systems analysis takes place in general. Several phases can be distinguished, with feedback possible from any phase to an earlier phase.

1. The first phase of systems analysis consists of understanding and translating into quantitative terms the objectives and performance requirements sought for the system in relation to the environment in which it operates.
2. The next phase is to formulate quantitatively (e.g., by a flow diagram) the structure and boundaries of the system.
3. Then a mathematical model has to be prepared for the system that includes all the possible interrelations between the variables that can be quantified. All quantifiable constraints must be included in the model, in addition to the functions yielding the input–output relations between the variables.
4. The coefficients in the model must be estimated and the desired input relations specified.
5. The model should be validated in light of the objectives established by step 1.

Figure 1.4 shows conceptually the cyclical nature of these phases.

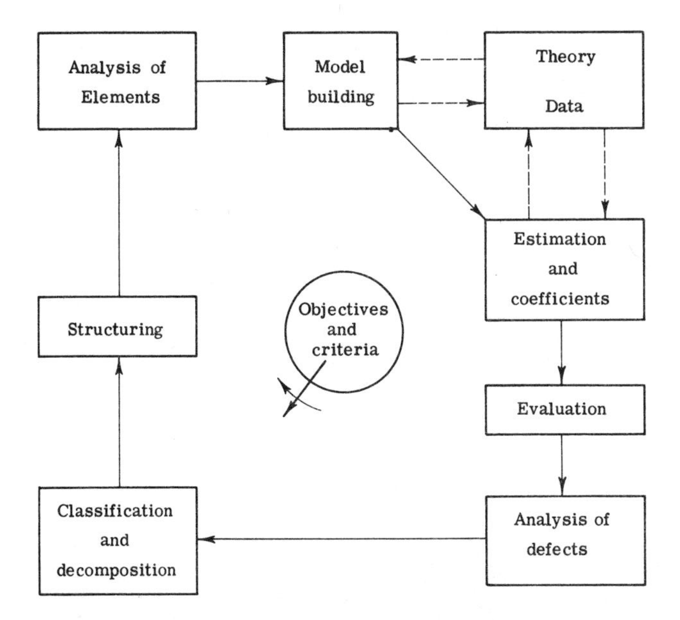

Fig. 1.4 Adaptive systems analysis.

Finally, after having taken into account each of the five phases in as much detail as required (but always being limited by the costs involved and by lack of information), the analyst is prepared to use the model for

1. *Economic experimentation.* The system time scale can be compressed by computer simulation so that existing and proposed operations can be examined in relatively reasonable times.
2. *Extrapolation.* Extreme ranges of operating conditions can be examined without incurring damages that might arise in a real physical system.
3. *Study of commutability and evaluation of alternate policies.* Elements of the system can be rearranged, new factors introduced, and the design of future systems evaluated.
4. *Effect of stochastic variables.* Random effects can be introduced with known statistics for the random variables.
5. *Sensitivity.* The effect on the outputs of changes in individual and joint variables and parameters can be examined.

All of the above uses of a model pertain to design and operation without actually undertaking the physical construction of the system.

In his work the systems analyst is particularly concerned with answers to the following questions:

1. For what organization or group of people is the system being developed?
2. What are the goals and objectives of the users of the system and other systems that interact with it?
3. What alternative systems should be considered?
4. What effects or consequences are imposed on the users by the various prospective systems?
5. What are the criteria for comparing and ranking alternative systems and for evaluating the consequences of building the system on objectives external to the system under consideration?
6. How reliable are the estimates of system costs and revenues, and of performance, and how does the level of reliability affect the choice of alternatives? How were the data used in the analysis obtained, and how reliable are they?
7. Are the consequences of the system alternatives measured, evaluated, and presented to the decision makers and to the community in which they function in a fashion that is understandable and usable?

The systems approach certainly opens the door for an expansion of analytical activity, and it involves a rather wide body of mathematical tools. It must also involve water resource management at the policy level, since most goals initially are specified in terms of somewhat generalized economic and social quantities rather than in specific physical terms.

We now turn to the question of how to relate the concepts of systems analysis to the problems of water resources management. There are two major areas of application: (1) the planning and (2) the operation of water resource systems. Planning for the unified development of a river basin consists of the collection of a data base followed by a series of decisions—e.g., when and whether to build each dam and canal, where to locate new towns and industries, how to operate the reservoirs, and so forth—as shown in Fig. 1.5. Planning is concerned with selecting from all possible alternatives that particular set of actions which will best accomplish *the overall objectives of the decision makers* [Hall and Dracup, 1970; Hillier and Lieberman, 1967].

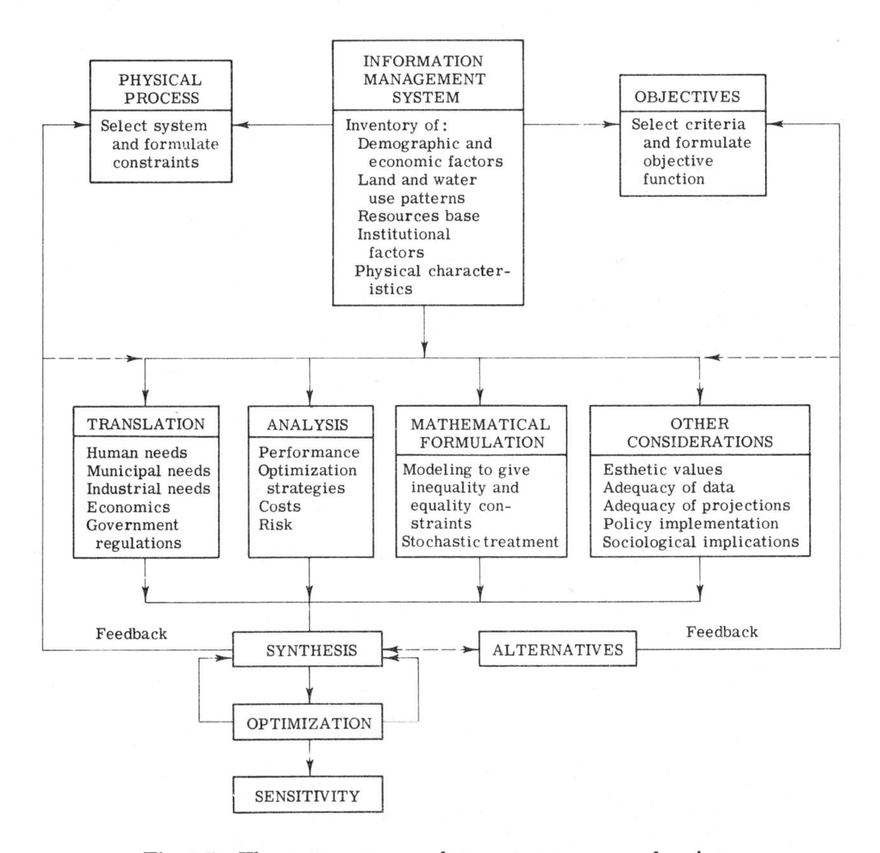

Fig. 1.5 The systems approach to water resources planning.

Operation of a water resources system, on the other hand, is concerned with what decisions are necessary to best accomplish the objectives of an existing system. While the operation of an existing water resources system may be considered disjointly from the planning function, the planning for the expansion of an existing system definitely must encompass the hypothesized future operation of the system. From the viewpoint of this book, *operation* is concerned with the optimization of an *existing system*, whereas *planning* attempts to formulate an optimal system by possible *additions of elements to the existing system.*

Because, as noted earlier, water resource development has changed in the last few decades from simple single-purpose projects to multipurpose programs involving large river basins, optimum planning, design, and operation can best be obtained through the use of mathematical models and high-speed digital computers. An analytical model, of course, incorporates many simplifying assumptions to make the model manageable. Even after many simplications, voluminous data are required to provide the coefficients and inputs for the mathematical model. Such data are often not available, nor are there adequate personnel to formulate and solve models. One must be wary that the models used do not become so simple that they no longer reflect the real physical system of the river basin. Some models are so rigid and mechanical that they cannot include the social benefits and costs of projects. Other models do not represent well the real river basin. Consequently, the analyst must be wary of attaching to a model a general aura of validity that it does not merit.

Once the five phases of systems analysis described above have been resolved, or at least partly resolved, two main routes can be pursued to realize the objectives for the system being planned, i.e., what is best in regard to the configuration of system elements and its operational policy. The two methods are simulation and optimization; see Fig. 1.6.

Simulation carries out "experiments" on a model of the system to obtain data that can be evaluated to determine the best operating policies [Ackoff, 1961; Hillier and Lieberman, 1967; Hufschmidt and Fiering, 1966]. Simulation was the first technique to be used by systems analysts in solving complex water resources problems. It was successful in examining the Nile Valley irrigation plan [Morrice and Allan, 1959], in planning for the augmentation of the Sydney, Australia water supply (1969), and in energy studies on the Columbia River [Lewis and Shoemaker, 1962]. The general procedure is to run case studies in which the operational parameters are varied for a number of preselected cases. Computer programs for a single case can be readily adapted to the case-study technique. One of the advantages of the case method is that it is not concerned with whether the

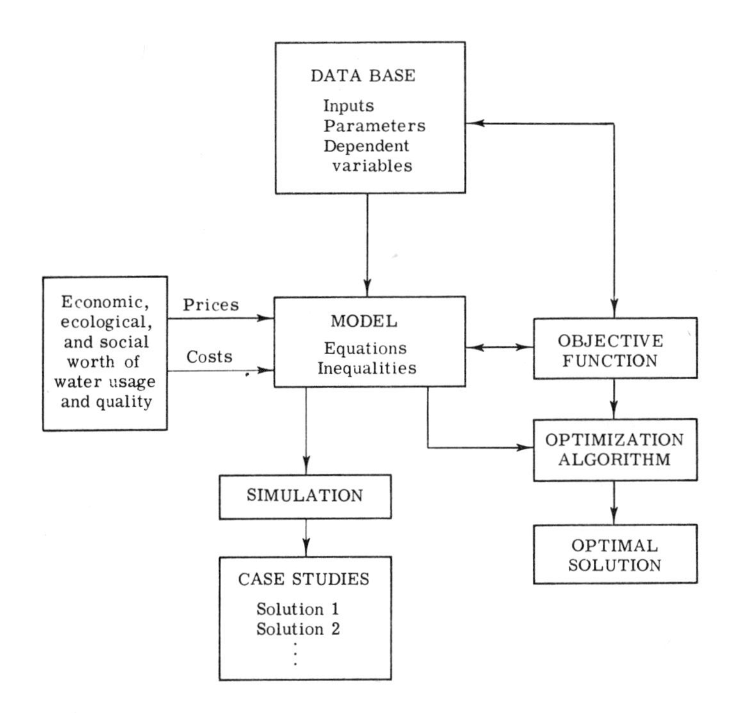

Fig. 1.6 Planning via simulation versus optimization.

functions in the mathematical model are continuous, because each case is treated as a separate entity. Simulation does not necessarily yield an optimal system and operating policy among all possible configurations; but if enough cases are run and the criteria used are not overly sensitive to changes in the system parameters, as is often the case, the best plan found among the cases will be a quite adequate plan. At present, simulation is used to view the outcome of those mathematical models (1) in which the variables are not deterministic but have a known statistical distribution, or (2) that are too complex to be handled by current optimization methods. We will be concerned only with modeling and optimization methods in this book i.e., the second of the two techniques.

To summarize, planning for the development of the water resources of a river basin requires examination of the following interrelated activities [Hall and Buras, 1961; Maass., 1962]:

1. the identification of objectives for the water resources system
2. the choice of the structure of the system design, i.e., the number and location of reservoirs, canals, lakes, and so on

3. the choice of the sequence of construction of system elements, i.e., when each dam and canal should be constructed and how large they should be

4. the choice of operating rules for the components of the river basin

Section 1.3 briefly describes some of the optimization techniques available for use in the water resources area, while Section 1.4 examines the criteria used for a water resources system.

1.3. Techniques for the Optimization of a Water Resources System

Optimization problems arise because rarely does a mathematical description of a water resources system yields exactly the proper number of independent equations to provide one and only one answer for the states (values of the dependent variables) in the model. A problem that admits of only one solution does not have to be optimized. The typical model is *underdetermined*; that is, there are fewer independent equations than there are variables whose values are sought. Such problems, in principle, have an infinite number of solutions; the objective of optimization is to select from the set of all possible solutions the best one(s) with respect to some given criteria.

Optimization can be accomplished by many strategies, ranging from quite sophisticated analytical and numerical mathematical procedures to the intelligent application of simple arithmetic. Assuming that the problem to be optimized is defined in some way, the two main methods of optimization can be conveniently classified as follows:

1. Analytical methods that make use of the classical techniques of differential calculus and the calculus of variations [Beveridge and Schechter, 1970, pp. 508–522, 618–625]. These methods seek the extremum of a revenue or *objective function* $f(\mathbf{x})$ by finding the values of $\mathbf{x} = [x_1, x_2, \ldots, x_n]^T$ that cause the derivatives of $f(\mathbf{x})$ with respect to $\mathbf{x}$ to vanish. When the extremum of $f(\mathbf{x})$ is sought in the presence of constraints, techniques such as Lagrange multipliers and constrained derivatives are used. For analytical methods to be used, the problem to be optimized must be described in a rather restricted way so that the functions and variables can be manipulated by known rules of mathematics. Analytical methods prove unsatisfactory for large, highly nonlinear problems, and will not be discussed in this text.

2. Numerical methods that generate solutions to the optimization problem by means of iterative procedures. Numerical methods can be used to solve problems that cannot be solved analytically. Because water resources problems prove tractable to numerical techniques, numerical methods are the ones to be considered here.

We will briefly summarize a few of the more important numerical optimization tools in this section in order to bring out the significant role that the model and objective function play in the optimization of a water resources system. The factors that can be included in a model of a water resources system and the form in which they must be included are dictated by the optimization methods that can be used. It is because of such restrictions that simulation has been used as an alternative to optimization, for simulation can accommodate almost any type of model.

1.3.1. Linear and Nonlinear Programming

One approach to optimization ignores the structure of the problem and on each iteration manipulates all the variables simultaneously in the quest for an optimum. This approach may be referred to as the simultaneous optimization technique, and is typified by linear and nonlinear programming. Linear programming [Beveridge and Schechter, 1970, pp. 287–324] has been used to solve such diverse problems as:

1. analysis of water resource decisions in international river basins [Rogers, 1969]
2. allocation of capital for water resources development [Marglin, 1962; Massé and Gibrat, 1957; Massé, 1962]
3. finding reservoir operating rules [Loucks, 1969; Manne, 1960; Thomas and Revelle, 1966]
4. treatment of polluted water [Lynn *et al.*, 1962; Revelle *et al.*, 1968; Sobel, 1965; Thomann and Sobel, 1964]

A linear programming problem is one in which a *linear* function is the criterion to be minimized or maximized, a criterion subject to constraints that are also *linear* functions. A combination of scalars or vectors denoted in general by X_i is said to be *linear* if the scalars or vectors can be assembled in the form

$$c_1X_1 + c_2X_2 + \cdots + c_nX_n$$

where the c's are constants. For example, the function

$$4x_1 + 3x_2 + 5x_3 + 2$$

is linear in the variables x_1, x_2, and x_3, whereas the function

$$2x_1^2 + x_1x_2 + 3\exp(x_3)$$

is nonlinear in the same variables.

Although the linear programming problem can be stated in many related forms, we will write it as follows [Dantzig, 1963; Wilde and Beightler, 1967]:

Maximize

$$f(\mathbf{x}) = \sum_{i=1}^{n} c_i x_i \tag{1.1a}$$

subject to

$$\sum_{i=1}^{n} a_{ij}x_i - b_j \leq 0 \qquad j = 1, \ldots, m \tag{1.1b}$$

$$x_i \geq 0 \qquad i = 1, \ldots, n \tag{1.1c}$$

where the a's, b's, and c's are constants and the x's are the variables whose values are sought. (If equality constraints are involved in the problem,

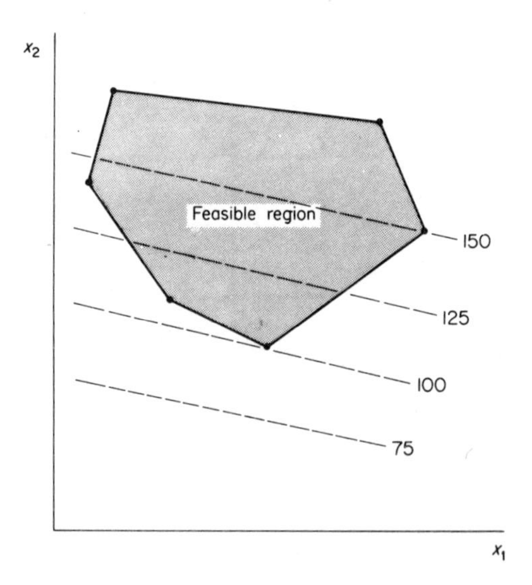

Fig. 1.7 The linear programming problem in two variables. The feasible region lies on or within solid lines representing the six constraints. A primal method of solution searches for a maximum among the vector of the vertices formed by the intersection of the constraints. Broken lines are contours of the objective function; solid lines are constraints $\sum_{i=1}^{n} a_{ij}x_i - b_j = 0$, $j = 1, \ldots, m$.

they can be changed into two inequality constraints, or, alternatively, used to reduce the vector of variables by one variable for each equation.) Figure 1.7 illustrates the linear programming problem for two variables. Various methods have been proposed to solve the problem posed by Eqs. (1.1), references for which can be found in the list of supplementary readings at the end of the book. Probably the best-known method is the revised simplex method.

Matrix notation provides a compact way of stating mathematical programming problems and describing algorithms for their solution. Let $\mathbf{x}$ and $\mathbf{c}$ be $n \times 1$ column vectors in E^n (i.e., in the n-dimensional Euclidean space composed of the n variables), $\mathbf{a}$ be an $n \times m$ matrix of constants, and $\mathbf{b}$ be an $m \times 1$ column vector:

$$\mathbf{x} = \begin{bmatrix} x_1 \\ x_2 \\ \vdots \\ x_n \end{bmatrix}, \qquad \mathbf{a} = \begin{bmatrix} a_{11} & a_{12} & \cdots & a_{1m} \\ a_{21} & a_{22} & \cdots & a_{2m} \\ \vdots & \vdots & \vdots & \vdots \\ a_{n1} & a_{n2} & \cdots & a_{nm} \end{bmatrix}, \qquad \mathbf{b} = \begin{bmatrix} b_1 \\ b_2 \\ \vdots \\ b_m \end{bmatrix}, \qquad \mathbf{c} = \begin{bmatrix} c_1 \\ c_2 \\ \vdots \\ c_n \end{bmatrix}$$

Then the equivalent of Eqs. (1.1) in matrix notation is

Maximize

$$f(\mathbf{x}) = \mathbf{c}^{\mathrm{T}}\mathbf{x} \tag{1.2a}$$

subject to

$$\mathbf{ax} \leq \mathbf{b} \tag{1.2b}$$

$$\mathbf{x} \geq \mathbf{0} \tag{1.2c}$$

where the superscript T denotes transpose. A vector $\mathbf{x}^*$ satisfying expressions (1.2) is the desired solution.

Associated with every linear programming problem is a related problem termed the "dual" [Beveridge and Schechter, 1970, pp. 325–346]:

Minimize

$$f(\mathbf{u}) = \mathbf{b}^{\mathrm{T}}\mathbf{u} \tag{1.3a}$$

subject to

$$\mathbf{a}^{\mathrm{T}}\mathbf{u} \geq \mathbf{c} \tag{1.3b}$$

$$\mathbf{u} \geq \mathbf{0} \tag{1.3c}$$

A certain symmetry exists between the dual and the original problem (called the "primal" problem). If the objective in the primal problem is to find the maximum, the dual problem pertains to finding a minimum, and vice versa. The variables in the dual problem usually relate to certain costs or prices (usually called sensitivity coefficients, or shadow prices) that are attributed to the resources and/or activities of the problem.

As a very simple example of planning for investment using linear programming, we can consider the optimal allocation of funds to meet projected power requirements by five methods: (1) thermal power stations, (2) hydroelectric stations with reservoirs, (3) hydroelectric stations on rivers, (4) power stations with sluice installations, and (5) power stations operated by means of ocean tidal basins. Table 1.2 presents the essential technical data to solve the problem. The elements of the first row in Table 1.2 merely indicate that each type of subsystem is a candidate. Let the total guaranteed capacity of the five types of subsystem be denoted by x_1 through x_5. Then the following inequalities have to be satisfied for the guaranteed capacity, peak capacity, and yearly output, respectively:

$$x_1 + x_2 + x_3 + x_4 + x_5 \geq 1692,$$

$$1.15x_1 + 1.20x_2 + 1.10x_3 + 3x_4 + 2.15x_5 \geq 2307$$

$$7x_1 + 1.30x_2 + 1.20x_3 + 7.35x_4 + 5.45x_5 \geq 7200$$

In addition

$$x_1 \geq 0, \quad x_2 \geq 0, \quad x_3 \geq 0, \quad x_4 \geq 0, \quad x_5 \geq 0$$

because negative numbers for the production of power are inadmissible. Various combinations of power stations are feasible within the specified bounds.

To complete the problem statement, an objective function has to be formed. A number of possible criteria exist. For example: the combined construction costs might be a minimum, that is, the capital investment would be a minimum. A second criterion might be a requirement that the annual operating expenses be a minimum. Lastly, for the efficiency of the

Table 1.2

Data for Linear Programming Example

		Type of power station					
		1	2	3	4	5	Units
Guaranteed capacity	a_i	1	1	1	1	1	MW
Peak capacity	b_i	1.15	1.20	1.10	3	2.15	MW
Yearly output	c_i	7	1.30	1.20	7.35	5.45	GW-hr
Current construction costs	k_i	24	32	105	77	80	10^6 \$
Yearly operating costs	s_i	14	10	5.6	14	7.9	10^6 \$

investment to be highlighted, the solution yielding a minimum of the total of the current joint construction costs plus the discounted sum of the operating expenses may be sought. Suppose we use the latter, discounted at 8%, as the criterion of optimality. The problem can then be reduced to one of finding the minimum of the following linear objective function:

$$f(\mathbf{x}) = \sum_{i=1}^{5} k_i x_i + \sum_{t=1}^{T_{max}} [1/(1+r)^t] \sum_{i=1}^{5} s_{it} x_i$$

where r is a discount factor, k_i is the capital investment for project i, s_{it} represents the operating costs for project i in year t, and T_{max} is the planning horizon of 4 yr. With $r = 0.08$, we obtain

$$f(\mathbf{x}) = (24 + (1/1.08)(14) + (1/1.08^2)(14) + \cdots)x_1 + \cdots$$

Application of a linear programming code to minimize $f(\mathbf{x})$ yields the solution

$$x_1 = 694, \quad x_2 = 825, \quad x_3 = 0, \quad x_4 = 0, \quad x_5 = 0$$

$$f(\mathbf{x}) = 1.24 \times 10^5$$

As is clear from problem (1.1), the major handicap in the use of linear programming is that all the functions and inequalities (and equations) must be linear and the variables continuous. If they are not, then nonlinear programming [Beveridge and Schechter, 1970, pp. 355–502] must be employed.

The nonlinear programming problem can be formally stated as

Minimize

$$f(\mathbf{x}) \quad \text{for} \quad \mathbf{x} \in E^n \tag{1.4a}$$

subject to m linear and/or nonlinear equality constraints

$$h_j(\mathbf{x}) = 0 \qquad j = 1, \ldots, m \tag{1.4b}$$

and $(p - m)$ linear and/or nonlinear inequality constraints

$$g_j(\mathbf{x}) \leq 0 \qquad j = m+1, \ldots, p \tag{1.4c}$$

Figure 1.8 illustrates the nonlinear programming problem for two variables. The feasible region is identified by the hash marks along the equality constraint. Although in some special cases the equality constraints can be solved explicitly for selected variables and those variables eliminated from the problem as independent variables, reducing the problem to one with inequality constraints only, most often the equality constraints can only be solved implicitly and must be retained.

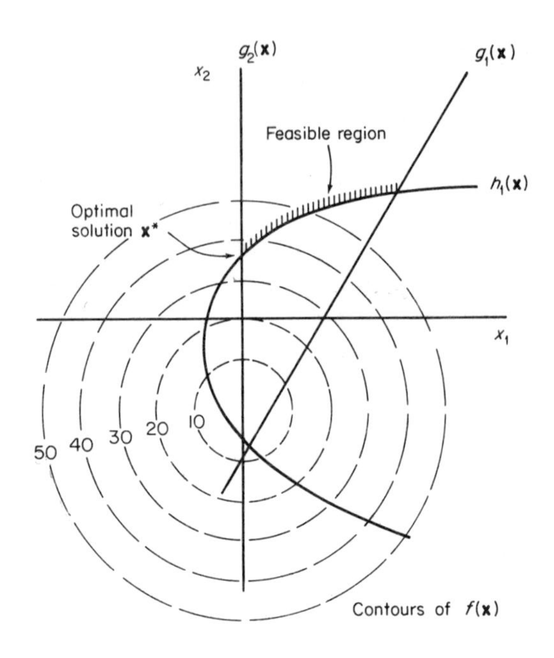

Fig. 1.8 Two-dimensional representation of the nonlinear programming problem.

An alternate representation of expressions (1.4a)–(1.4c) sometimes encountered is

Minimize

$$\{f(\mathbf{x}) \mid \mathbf{x} \in R\} \tag{1.5}$$

where R is the domain of $\mathbf{x}$ for which conditions (1.4b) and (1.4c) are satisfied, i.e.,

$$R = \{\mathbf{x} \mid h_j(\mathbf{x}) = 0; \quad g_j(\mathbf{x}) \leq 0 \quad \text{for all } j\} \tag{1.6}$$

The inequality sign in $g_j(\mathbf{x}) \leq 0$ can be reversed by multiplying through by -1 without changing the correctness of the statement of the problem.

Nonlinear programming has been used to determine the optimal operation of reservoirs [Lee and Waziruddin, 1970] and the least cost of a mix of reservoirs, wells, desalinization, and treated waste water for the James River region [Young and Pisano, 1970]. A number of solution techniques are available to solve the nonlinear programming problem (see Himmelblau [1972] or Wilde and Beightler [1967, Chapter 2]).

To provide a brief example of a nonlinear programming problem, consider the operational policy for three reservoirs in series as illustrated in

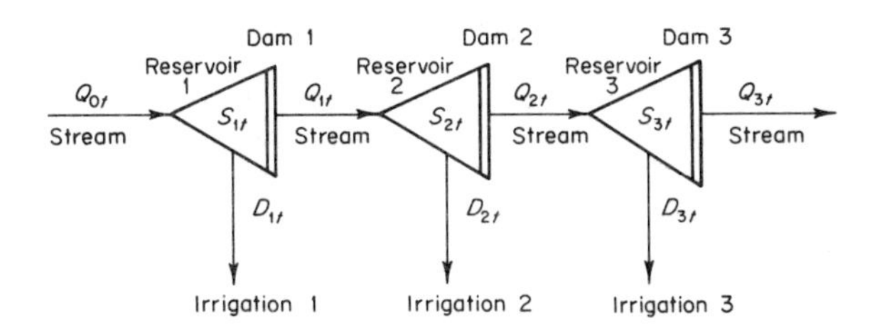

Fig. 1.9 Three-reservoir system.

Fig. 1.9. Using units for the notation as given in Appendix C, let S_{jt} be the storage volume of water in reservoir j at the beginning of period t, D_{jt} the amount of water supplied for irrigation from reservoir j in period t, and Q_{mt} the flow in stream m during period t—all in acre-feet. The mass balances on the water flow are

$$\begin{array}{ccccc} \text{accumulation} & & \text{input} & & \text{output} \\ S_{1,t+1} - S_{1t} & = & Q_{0t} & - & Q_{1t} - D_{1t} \\ S_{2,t+1} - S_{2t} & = & Q_{1t} & - & Q_{2t} - D_{2t} \\ S_{3,t+1} - S_{3t} & = & Q_{2t} & - & Q_{3t} - D_{3t} \end{array}$$

and the initial conditions are S_{j0}. Inequality constraints come into play because releases to the streams between the reservoirs must always exceed certain minimum levels for downstream usage but must not exceed flood levels:

$$Q_{mt}^{\min} \leq Q_{mt} \leq Q_{mt}^{\max} \qquad m = 1, 2, 3, \quad t = 1, \ldots, T_{\max}$$

Similar upper and lower bound exist on the capacity of the irrigation canals:

$$D_{jt}^{\min} \leq D_{jt} \leq D_{jt}^{\max} \qquad j = 1, 2, 3, \quad t = 1, \ldots, T_{\max}$$

A third set of constraints pertains to the storage capacity and minimum level in each reservoir:

$$S_{jt}^{\min} \leq S_{jt} \leq S_{jt}^{\max} \qquad j = 1, 2, 2, \quad t = 1, \ldots, T_{\max}$$

Keep in mind that the upper and lower bounds are known constants. To put the above bounds in the form of inequality (1.1b), the lower bounds for Q_{mt} are $Q_{mt}^{\min} - Q_{mt} \leq 0$ and the upper bounds are $Q_{mt} - Q_{mt}^{\max} \leq 0$; the other bounds can be similarly rearranged.

Table 1.3

Data for Nonlinear Programming Example

Reservoir	S_{m0} (acre-ft)	D_{jt}^{max} (acre-ft)[a]	S_{jt}^{max} (acre-ft)[b]	Q_{mt}^{min} (acre-ft)	Q_{mt}^{max} (acre-ft)	E_{jt}
1	83,400	3,000	95,000	400	8,000	85,000
2	87,560	3,000	95,000	400	8,000	89,000
3	92,500	3,000	95,000	400	8,000	93,000

Period, t	Q_{0t} (acre-ft)
1	1800
2	3100
3	3600
4	3200
5	2000

Values of the coefficients in the objective function including the discount factor

Variable	Period, t: 1	2	3	4	5
a_{1t}	0.65×10^{-3}	0.83×10^{-3}	0.90×10^{-3}	0.83×10^{-3}	0.65×10^{-3}
a_{2t}	0.43×10^{-4}	0.54×10^{-4}	0.51×10^{-4}	0.45×10^{-4}	0.32×10^{-4}
a_{3t}	0.78×10^{-4}	0.10×10^{-3}	0.11×10^{-3}	0.10×10^{-3}	0.78×10^{-4}
b_{1t}	0.23×10^{-1}	0.24×10^{-1}	0.24×10^{-1}	0.24×10^{-1}	0.23×10^{-1}
b_{2t}	0.22×10^{-1}	0.23×10^{-1}	0.24×10^{-1}	0.23×10^{-1}	0.22×10^{-1}
b_{3t}	0.47×10^{-1}	0.47×10^{-1}	0.50×10^{-1}	0.47×10^{-1}	0.47×10^{-1}
c_{1t}	0.39×10^{3}	0.35×10^{3}	0.28×10^{3}	0.35×10^{3}	0.39×10^{3}
c_{2t}	0.58×10^{3}	0.52×10^{3}	0.42×10^{3}	0.52×10^{3}	0.58×10^{3}
c_{3t}	0.13×10^{4}	0.12×10^{4}	0.97×10^{3}	0.12×10^{4}	0.13×10^{4}
d_{1t}	0.87×10^{-5}	0.78×10^{-5}	0.64×10^{-5}	0.78×10^{-5}	0.87×10^{-5}
d_{2t}	0.82×10^{-5}	0.74×10^{-5}	0.60×10^{-5}	0.74×10^{-5}	0.82×10^{-5}
d_{3t}	0.76×10^{-5}	0.68×10^{-5}	0.56×10^{-5}	0.68×10^{-5}	0.76×10^{-5}

[a] D_{jt}^{min} is zero.
[b] $S_{jt}^{min} = 300$.

As to a nonlinear objective function, let us assume that the irrigation revenues are a quadratic function of D_{jt}:

$$\text{irrigation revenue} = a_{jt}D_{jt}^2 + b_{jt}D_{jt}$$

where a_{jt} and b_{jt} are known constants. Let us also suppose that recreational benefits can be expressed by

$$\text{recreation revenue} = c_{jt} - d_{jt}[E_{jt} - S_{jt}]^2$$

where c_{jt} and d_{jt} are known constants and E_{jt} is the optimal amount of water in the reservoir. Then the objective function to be maximized would be

$$f(\mathbf{x}) = \sum_{t=1}^{T_{\max}} \sum_{j=1}^{3} \alpha\{a_{jt}D_{jt}^2 + b_{jt}D_{jt} + c_{jt} - d_{jt}[E_{jt} - S_{jt}]^2\}$$

where α is the discount factor that reduces all dollars to their present value. Table 1.3 lists typical values from Lee and Waziruddin [1970] for all the constants for the above nonlinear programming problem, and Table 1.4 gives the best solution they obtained for $T_{\max} = 5$ (for a constraint error of less than 5 acre-ft) by both the conjugate gradient and gradient

Table 1.4

Solution of the Nonlinear Programming Problem[a]

Initial guess for the independent variables for period $t = 1, \ldots, 5$

j	D_{jt}	Q_{jt}
1	2500	1000
2	1000	2000
3	2500	1000

Solution

Value of objective function $f(\mathbf{x}) = 47{,}430$.
Values of Q_{jt} were as follows:

	t				
j	1	2	3	4	5
1	400	500	1200	1300	1000
2	1700	1800	2300	2100	2200
3	400	400	400	400	400

Values of D_{jt} were $D_{1t} = D_{3t} = 3000$, and $D_{2t} = 0$, $t = 1, \ldots, 5$.

[a] Based on the data in Table 1.3.

projection algorithms. For $T_{max} = 5$ the problem comprises 45 variables, 15 equality constraints, and 90 inequality constraints.

1.3.2. Dynamic Programming

An important feature of most mathematical models of real systems is that even if all the subsystems are optimized in separate phases, the whole system is not necessarily optimal. However, a water resources system has a unique characteristic, namely that all the water in the river basin flows downhill! Therefore, unless there is recycling of water back upstream, say by pumping, another class of optimization techniques is applicable—those that make use of the characteristic information flow in the system, such as dynamic programming [Beveridge and Schechter, 1970, pp. 679–702]. Dynamic programming is able to exploit the structure of a problem by decomposing it into sequence of optimal subproblems, each of which is of a smaller scale than the original problem. Hall and Buras [1961] first pointed out the efficacy of dynamic programming in finding operating rules for reservoirs. Since then, much work on refining operating rules using dynamic programming has been reported in the literature [Amir, 1967; Butcher, 1968; Hall, 1964; Hall *et al.*, 1968; Hall and Howell, 1963; Mobasheri and Harboe, 1970; Schwerg and Cole, 1968; Young, 1967].

Dynamic programming proceeds to optimize the elements of a system in the inverse direction to the information flow in the system. For example, in Fig. 1.9 reservoir 3 would be optimized first, then reservoirs 2 plus 3, and finally 1 plus 2 plus 3. Note that whatever the choice will be for the irrigation water for reservoir 3 in the optimization of reservoir 3, the choice cannot affect the optimization of any of the upstream reservoirs because of the information flow. Hence D_{3t} is chosen to optimize the return from reservoir 3, D_{2t} can be selected to optimize the return from reservoirs 1 plus 2 independently of the choice of D_{3t}, and so on. Although dynamic programming can be used to handle problems with nonlinear objective functions and constraints, if too many (more than two) decision variables such as D_{jt} exist at each stage, the degree of effort required for the search for the optimum at each stage becomes excessive.

The dynamic programming problem can be formally written as follows. For each stage of the serial system there is a nonlinear difference equation that represents the "state" equation

$$\mathbf{x}_{t+1} = \boldsymbol{\phi}_1(\mathbf{x}_t, \mathbf{y}_t, t) \tag{1.7}$$

where $\mathbf{x}_t$ is the n-dimensional state (dependent) variable vector comprising $j = 1, \ldots, n$ stages, $\mathbf{y}_t$ the n-dimensional decision (independent, control) vector comprising $j = 1, \ldots, m$ stages, $\boldsymbol{\phi}_1$ the n-dimensional functional, and $\mathbf{x}_0$ a constant initial-condition vector. In addition there is an objective function

$$f(\mathbf{x}, \mathbf{y}) = \sum_{t=1}^{T_{\max}} \phi_2(\mathbf{x}_t, \mathbf{y}_t, t) \tag{1.8}$$

and some inequality constraints that define the feasible region for search,

$$\mathbf{x}_t^{\min} \leq \mathbf{x}_t \leq \mathbf{x}_t^{\max} \qquad \mathbf{y}_t^{\min} \leq \mathbf{y}_t \leq \mathbf{y}_t^{\max} \tag{1.9}$$

The problem is to determine $\mathbf{y}_0, \ldots, \mathbf{y}_{T_{\max}}$ so that $f(\mathbf{x})$ is maximized subject to relations (1.7) and (1.9). At any stage $(k + 1)$ the optimum for that stage is the optimal return from all the downstream (in time or space) stages plus the return for the $(k + 1)$st stage:

$$f(\mathbf{x}, \mathbf{y}) = \max_{\mathbf{y}_{k+1}} \{f_{k+1}(\mathbf{x}, \mathbf{y}, t) + \max \sum_{t=k+2}^{T_{\max}} f_t(\mathbf{x}, \mathbf{y}, t)\} \tag{1.10}$$

As a very simple numerical example, consider the one-period optimization problem of the operation of three reservoirs in series as shown in Fig. 1.9. Storage and release are for one time period t. Minimum and maximum levels corresponding to expressions (1.9) are specified for the river flows $Q_{jt}, j = 1, 2, 3,$

$$4 \leq Q_{1t} \leq 12 \qquad 4 \leq Q_{2t} \leq 12 \qquad 4 \leq Q_{3t} \leq 12$$

and the water for irrigation D_j

$$0 \leq D_{1t} \leq 12 \qquad 0 \leq D_{2t} \leq 12 \qquad 0 \leq D_{3t} \leq 12$$

The system equations corresponding to Eq. (1.7) are linear equality constraints that represent the material balances on the water flows for one period Δt if there are no changes in reservoir levels, i.e., $S_{j,t+1} - S_{jt} = 0$ because the problem is for one period only:

$$Q_{1t} = Q_{0t} - D_{1t} \qquad Q_{2t} = Q_{1t} - D_{2t} \qquad Q_{3t} = Q_{2t} - D_{3t}$$

Note that the irrigation water releases might become delayed returns to the next reservoir, so that the irrigation water released during the period Δt might be D_{jt} leaving the jth reservoir but $\gamma D_{j,t-k}$ entering the $(j + 1)$st reservoir, where k represents a delay of one or more periods and γ is the fraction of water retained.

One of the advantages of dynamic programming is that the objective function does not have to be an analytical objective function nor contain continuous variables. Suppose we let the objective function be a table of data giving the values in thousands of dollars of the irrigation water withdrawn by each reservoir for a single time period t; we want to maximize the irrigation revenue. Note that the reservoir releases have been listed in multiples of four units for simplicity. Finally, assume that $Q_{0t} = 12$ (all flows are in thousand acre-feet).

	Quantity of water withdrawn, D_{jt} (10^3 acre-ft)			
Reservoir, j	0	4	8	12
	Objective function data, $f_t(D_{jt})$ (10^3 dollars)			
1	0	2	3	3
2	0	2	3	4
3	0	1	2	3

Starting with reservoir 3 we know the value of the objective function for reservoir 3, $f_3(D_{3t})$, as a function of the decision variable D_{3t} by reading across the bottom row of the tabular revenue. The third inequality constraint for river flow indicates that $Q_{3t} = 4$ (because if $Q_{3t} > 4$, no revenue will accrue from the flow greater than 4 and the excess water will be "wasted"), and the equality constraints give relations between D_{jt} and $Q_{j-1,t}$ such as

$$D_{3t} = Q_{2t} - 4 \qquad Q_{2t} = Q_{1t} - D_{2t}$$

Next we determine the revenue from reservoirs 2 and 3 combined as a function of D_{2t}. The flow to reservoir 2 can be 12, 8, or 4, depending on the upstream withdrawal, so that three cases must be examined for feasible solutions:

$Q_{1t} = 12$

D_{2t}	f_2	$\max f_3$	$f_2 + \max f_3$
0	0	2	2
4	2	1	3†
8	3	0	3†

$Q_{1t} = 8$

D_{2t}	f_2	$\max f_3$	$f_2 + \max f_3$
0	0	1	1
4	2	0	2†

$Q_{1t} = 4$

D_{2t}	f_2	$\max f_3$	$f_2 + \max f_3$
0	0	0	0†

The maximum revenue is designated by †. We note from the analysis that

$$\max f_{2+3} = \max[f_2(D_{2t}, Q_{1t}) + \max f_3(D_{3t}, Q_{2t})]$$

Finally we maximize all three reservoirs together such that

$$\max f_{1+2+3} = \max[f_1(D_{1t}, Q_{0t}) + \max f_{2+3}]$$

Because the input to reservoir 1 is given as $Q_{0t} = 12$, an analysis yields

$$Q_{0t} = 12$$

D_{1t}	f_1	$\max f_{2+3}$	$f_1 + \max f_{2+3}$
0	0	3	3
4	2	2	4†
8	3	0	3

The values of $\max f_{2+3}$ can be taken from the appropriate schedule under each Q_{1t}, i.e., $Q_{1t} = 12, 8$, or 4; they are marked by †. For the maximum revenue of 4, the allocation of irrigation water is 4 from reservoir 1, 4 from reservoir 2, and 0 from reservoir 3. Although this example has been a trivial one, it does illustrate many of the important features of dynamic programming, especially the substitution of several single-dimensional searches for one multidimensional search, and the variety of functions that can be accommodated.

1.4. Criteria for Optimal Planning

The objectives of any plan are to economically meet the current and future water needs of a given area. These needs (particularly in the agricultural sector) are subject to varying interpretations and often fail to meet minimum acceptable economic and ecological standards. It is certainly true that goals other than economic efficiency have influenced the decisions of policy makers in the field of water resources development. One has only to view the many projects and programs already completed that were intended primarily to stimulate and uplift the economy of a depressed region or to subsidize some sector of the economy. Nevertheless, our view will be a quantitative one, namely that all revenues (benefits) and costs can be explicitly expressed in terms of monetary prices.

Only rarely does the examination of the needs and goals for a river basin as stated verbally lead to a single quantitative criterion. If several separate criteria are to be used in evaluating what is "best," the analyst must some-

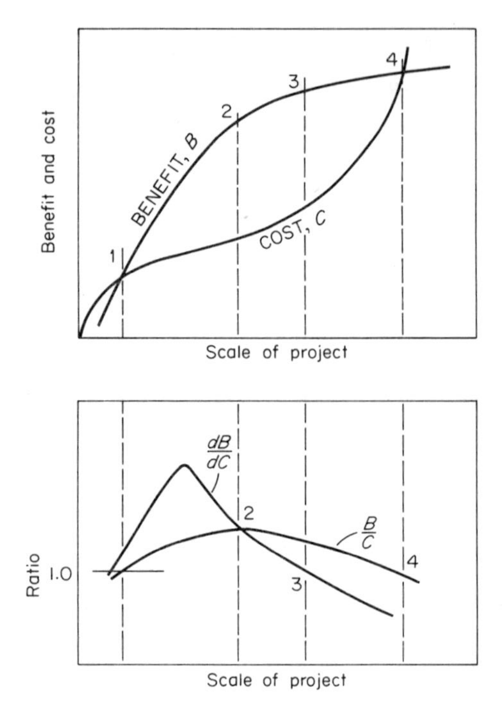

Fig. 1.10 Relationships between present value of benefits and cost for varying scales of project.

how weight the respective criteria or indicate the general domain of acceptability for each. The general mathematical expression(s) that details the revenue and costs contributed to the selected criterion is known as the *objective function.*

In determining the pertinent revenues and costs it is necessary to examine each physical quantity in the river basin together with the intangible revenues and costs, insofar as the latter can be quantified. Figure 1.10 illustrates figuratively the benefits and costs of a river basin project as a function of the scale of the project. Let us look at the costs first. One must be careful to avoid omision of important and costly features of the water resources system, such as drainage or leveling of the newly irrigated land, that may escape being included in the cost estimate. The cost of adverse effects such as relocation of transportation systems or resettlement of the population of an inundated area should be included in project cost, whether or not these costs are compensated. The project cost should include all the costs for the establishment, maintenance, and operation of the project. In a few instances costs will be subtracted directly from benefits to achieve

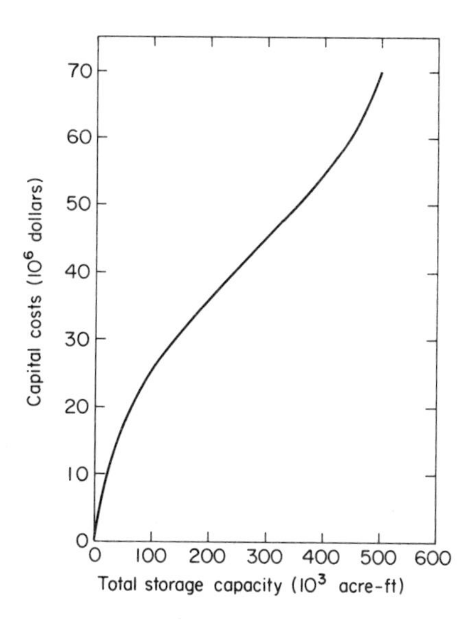

Fig. 1.11 Capital cost of a reservoir at current prices as a function of the total storage capacity. (From Water Resources Center [1968].)

net costs (or benefits). Figures 1.11–1.13 show the relation between capital costs and appropriate variables for dams, power plants, and irrigation.

Revenues (or benefits) arising mainly from flood control, new irrigation water, hydroelectric energy, domestic and industrial water supply, and so on are identified. The magnitude of these products and services is a direct function of the scale of the development, which in turn is a function of

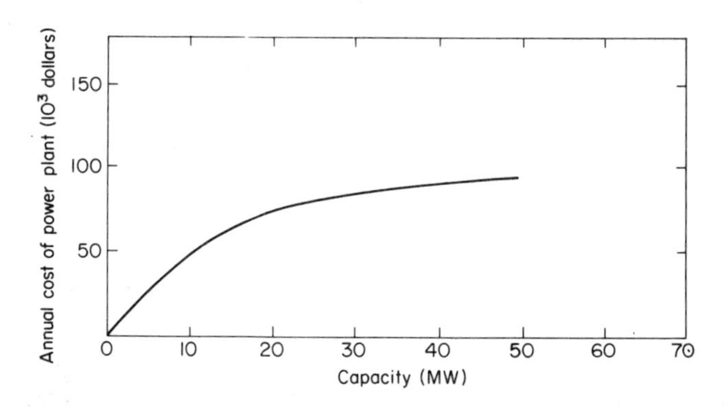

Fig. 1.12 Annual power plant cost at current prices as a function of capacity. (From Water Resources Center [1968].)

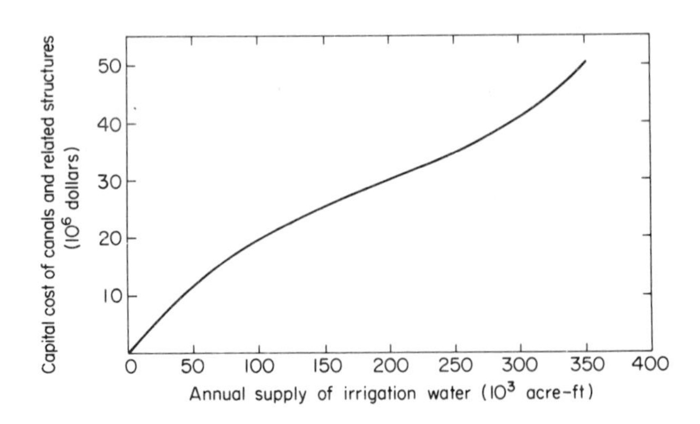

Fig. 1.13 Capital cost of irrigation canals and related structures at current market prices versus the quantity of irrigation water supplied. (From Water Resources Center [1968].)

costs incurred. The products or services are usually of two types: (1) increases in the production of valuable goods, such as energy, water, and so on, and (2) reductions in costs, such as flood damage.

Another category of benefits arising from the development of water resources has been called indirect or secondary benefits. These are defined as the increase in net incomes as a result of activities stemming from or induced by the project. Considerable controversy exists with regard to the inclusion of secondary benefits in the analysis. It has been argued that the available methods of measuring secondary benefits are such that they offer ample opportunities for abuse. Because these secondary benefits can be exaggerated to make the project appear to be economically feasible, some economists have recommended that secondary benefits should be dropped from the analysis [Ciriacy-Wantrup, 1955]. Others would agree that secondary benefits may be significant from a local or regional point of view but believe that from a national point of view they are merely transfer items from one region to another and therefore should not be included in the analysis [U.S. Government, 1958, p. 10].

What prices should be ascribed to the benefits and costs associated with the plan? One may question the implicit assumption that market prices correctly reflect the social value of benefits produced and productive factors employed. Nevertheless, alternative choices for pricing appear to be even less satisfactory, so that

> Despite the limitation of the market price system in reflecting values from a public viewpoint, there is no other suitable framework for evaluating the effects of public works projects in common terms. [U.S. Government, 1958, p. 8]

If market prices exist to measure the benefits of flood control, irrigation water, electric energy, and so on, then the market prices should be used to estimate benefits. In the absence of a market price, the estimated cost of the least costly alternative source that could be developed if the project under consideration is not put into effect might be used as a basis for benefit measurement. This method is used extensively to calculate the benefits from hydroelectric energy. The cheapest alternative, such as thermal or nuclear power plants, that could provide comparable electric energy would be used as the energy benefit from the development of the hydro project. It should be noted, however, that this alternative-cost computation will not be a correct measure of the benefits unless there is a real demand for the energy produced and the market can absorb the electricity made available. To calculate agricultural benefits or irrigation water, the concept of willingness to pay might be used if somewhat competitive markets are at work. An alternative approach is to use the "with and without" principle to estimate the agricultural benefits of a project; that is, the irrigation benefits of the project are calculated by the difference between the rural income with the project and without it.

A final consideration in establishing the criteria to be used once the prices are established is how to reduce the monetary values attached to the benefits and costs to a common point in time (the present) so that an evaluation of a project can be made. The usual method is to calculate the present value of benefits and costs using a discounting factor. If the value of benefit b at the period j is b_j, then its present value would be $b_j/(1 + r)^j$, where r is the discount or interest rate. Considerable controversy exists as to the appropriate discount rate.‡ Most federal agencies in the United States use the yield on the long-term federal bonds as the cost of capital and discount rate. If the market interest rate is employed as the correct measure of the cost of capital, when a high interest rate in the market must be used for the evaluation of projects, many of the projects will have a negative present value for their net return.

The problem with regard to the level of the interest rate becomes very significant in underdeveloped countries. Most of their public projects are large scale and relatively capital-intensive, and have long economic life. On the other hand, because of the scarcity of investment capital, interest rates are very high—much higher than in the more developed countries. The interest rate is even higher in the unorganized money market for the majority of the underdeveloped countries where the weighted average rate of interest is somewhere between 24 and 36% per annum [Meier, 1964].

‡ See, for example, the *Fed. Regist.* (1971) and de Neufville and Stafford (1971).

Even if an interest rate of 10% is used in the evaluation of public projects, the chances are slim that many projects in the field of water resources development can pass an economic feasibility test.

1.5. Previous Work in Optimal Planning for a Water Resources System

Most work that has been reported in the last two decades has been devoted to finding the best reservoir operating rules (refer to Section 1.3). However, in the last few years some attention has been focused on finding the optimal sizing and time of construction of system elements [Hufschmidt, 1962; Generoso, 1966; Wallace, 1966; McLaughlin, 1967; Howard and Nemhauser, 1968; Butcher *et al.*, 1969; Young *et al.*, 1969; Woolsey, 1969; Weiss and Beard, 1970; Hinomoto, 1970; Morin, 1970; Nayak and Arora, 1970]. Hall and Shepard [1967] used a combination of linear and dynamic programming to find the reservoir operating rules of a complex river system comprising the rivers, canals, and dams of Northern California that were part of the California Water Plan. Moseley *et al.* [1969], Young *et al.* [1970], and Evenson and Moseley [1970] examined the necessary dimensions and sequence of construction of reservoirs and canals for the Texas Water Plan [Texas Water Development Board, 1968].

Orlob [1970] described the approach taken by the planners for the Texas Water System. As shown in Fig. 1.1, there would be 18 reservoirs and more than 500 miles of canals; in addition there would be pumping facilities to raise the water from sea level to over 3000 ft elevation. When posed as a planning problem, the problem stated in words is

Given: (1) the location of all the reservoirs, (2) the roots of the interconnecting canals, (3) schedules for the in-basin demand for each reservoir and each major junction of the system, (4) the hydrology of supply for each major storage element, (5) the cost of imported water, and (6) the costs of construction, operation, and maintenance for all the elements;

Find: the least costly alternative system and schedule for its construction to meet the specified demands to the year 2020 within the prescribed legal, financial, contractual, and political constraints.

The group used a combination of linear programming, simulation, response surface methods, and perturbation analysis. Their approach was to seek "near optimum" solutions rather than exact optima to overcome the limits that existed on computation time and computer facilities.

The work of Hufschmidt, Wallace, Young, Orlob, and Woolsey will be briefly summarized in this section because these authors illustrate the main approaches used so far in solving the problem of the sequencing of construction of system elements and the increasingly sophisticated problems that may be solved.

One of the most extensive works in the last two decades treating water resources systems including their optimization was carried out by the Harvard Water Program [Maass, 1962]. This study introduced the objectives, defined the concepts, improved the methodology, and pointed out the factors that were relevant to improving water resources systems. For the first time, both simulation and optimization techniques were combined to find the best policies. As part of the Harvard study, Hufschmidt carried out the first systematic study of the optimal sizing of reservoirs in a water resources system designed to maximize the return on investment and to meet a schedule of water demands. He assumed (1) a configuration of dams and rivers, (2) an operating rule for each dam (not necessarily the "optimal" one), (3) a deterministic hydrology of sixty years, and (4) that the demand schedule did not vary from year to year. For each combination of dam sizes a simulation gave the revenue from operating the system as well as any irrigation water and energy shortages incurred. The strategy of searching for the optimum was as follows:

1. By random sampling over the independent variables, reduce the range of the variables that have to be searched. The probability that the combination of dam sizes having the highest revenue lies within a fixed percentage of the optimum can also be calculated [Young *et al.*, 1969].
2. Pick a "likely" starting point from within the reduced ranges for search. Then use a gradient search technique to find the optimum point.
3. In the region of the optimum point, use a grid sampling technique and/or "marginal" (incremental) analysis to ensure that the point is a "local" optimum.
4. Select other "likely" points from which the optimum solution can be obtained. If the response function is not convex, the gradient search technique cannot guarantee that the local optimal point has been found. If the same optimum point is located from several disparate starting points, one can be reasonably sure that a reasonable optimum point has been found.

Wallace [1966] advanced the work of Hufschmidt in certain respects. He used linear programming to plan the sizing of new reservoirs to meet projected power and irrigation demands in the Maule River (in Central Chile) and to provide the reservoir operating rules.

Young *et al.*, [1969] and Orlob [1970] looked at the policy of the sequencing of reservoir construction to meet increasing water demands over time. They took as the physical system the configuration of dams in East Texas required for annual transportation of 12 million acre-ft of water in Texas, as specified by the Texas Water Plan [Texas Water Development Board, 1968]. They examined a prespecified ultimate network configuration and assumed a deterministic hydrology. There were two reservoir capacities at each site, either zero or the design capacity; that is, a reservoir was either not built or it was built. For each combination of dam sizes a simulation (similar to network analysis) gave the return from operating the system to minimize pumping and maintenance costs.

The method of analysis was similar to Hufschmidt's and can be summarized as follows.

1. Preliminary sizes of elements of the system and operating rules for the reservoir were determined by a formal optimization procedure.
2. An initial screening was carried out by simulation of the given hydrology, element sizes, and operating rules for a large number of alternative development schedules selected by random sampling of the cost response surface. The range of variables was reduced by random sampling over the independent variables.
3. A gradient search was used to further reduce the range of variables to be searched. (The surface of the objective function used contained many crevices and was not concave.)
4. The most attractive schedules were improved by a method of successive perturbations.
5. Element sizes were further refined by a second simulation procedure, which constrained the flows in some of the expensive canals.
6. A second screening was carried out via a formal optimization of the most attractive systems and development schedules.
7. Finally, a pattern search [Hooke and Jeeves, 1961] was used to reach the optimum in the vicinity of the crevices.

Woolsey [1969] looked at the problem of competing public and private investments in water resources, a problem first formulated quantitatively by Steiner [1959], as applied to an actual problem in the Delaware River Basin. Because the problem was formulated as an integer programming problem, the method of solution used was the partial enumeration algorithm of Balas [1965], as modified by Glover [Glover and Zionts, 1965], and coded by Peterson [1967]. The model of the river basin allowed for (1) a choice of alternative dam sizes at each site, and (2) a way to satisfy the energy and irrigation demands by operating several dams in concert.

It did not allow for the spatial configuration of the system. The solution found was the optimal one.

1.6. Summary

From the foregoing discussion one can conclude that resolving the question of the sequencing of capital investment in water resources projects and the choice of reservoir operating rules can be approached by several methods. The problems that have been solved by simulation and optimization have grown more complex over the years because of (1) the increased sophistication of the techniques available, (2) the increased familiarity of systems analysts with these techniques, and (3) the increased efficiency of the computers available for implementing the necessary calculations. The next two chapters treat the problem of (1) the timing and amount of the capital investment, (2) the spatial configuration of the system elements, and (3) the reservoir operating rules, by first formulating a problem in quantitative terms that is compatible with the solution techniques recommended to solve the problem, and then describing the algorithms that can be applied to resolving the problem.

Chapter 2

FORMULATING THE PROBLEM OF THE OPTIMAL EXPANSION OF AN EXISTING WATER RESOURCES SYSTEM

In this chapter the problem of expanding an existing water resources system is formulated in mathematical terms by providing functions, equations, and inequalities that represent appropriate criteria and the characteristics of the physical system. The locations of possible dam sites are assumed to be determined by the topography and runoff of a particular region [McLaughlin, 1967]. Of interest here is the question of when new elements should be added to an existing system and how large the new elements (such as dams and canals) should be. We shall first consider some of the major assumptions and decisions that must be made in formulating the problem statement, and then take up in order (1) establishment of an objective function representing the criteria for the river basin, and (2) formulation of a model of the river basin, including identification of the different elements and the costs and benefits.

2.1. Assumptions Made in Preparing the Problem Statement

To prepare a problem statement that is compatible with the solution techniques available, certain assumptions must be made with respect to water resources development. The most significant assumptions are as follows.

1. The time scale for the introduction of new projects is chosen to be a year. Each new project becomes part of the system at the start of the year. Planning capital investment projects on a yearly time base is the accepted planning procedure in both the public and private sectors [Lesso,

1967; Butcher *et al.*, 1969; Young *et al.*, 1969; Weiss and Beard, 1970.] On the other hand, controlled water releases within the system are executed each month because (a) hydrologic data are provided on a monthly basis by such federal agencies as the U.S. Bureau of Reclamation and the U.S. Geological Survey, and (b) demands (consumptive and nonconsumptive) on the system are specified month by month. The monthly time frame for reservoir operation has been widely used by other investigators [Hufschmidt, 1962; Wallace, 1966; McLaughlin, 1967; Young *et al.*, 1969; Hall and Dracup, 1970].

2. As presently formulated, the model of the river basin does not take into account flood damages. Since these damages can be correlated with average monthly flows [Hufschmidt, 1962; Hufschmidt and Fiering, 1966], the model can be modified to rectify this omission if desired.

3. Reservoir inflows over the life of the project are assumed to be known and deterministic. In arid and semiarid areas the record of reservoir inputs will contain, in many instances, a sequence of subnormal flows called the "critical period." This critical-period hydrology has been used in many instances for the evaluation of optimal long-term policies for planning [Hall and Dracup, 1970].

4. It has been assumed that future demands on the system are deterministic. However, the mathematical model can incorporate any likely demand sequence, and the sensitivity of decisions to changes in the demand sequence can be ascertained.

5. Water quality has been omitted in forming the model because the model pertains to the quantity of water only; thus the surface water in a river system has been treated as a homogeneous commodity. How to implement the basic model to accommodate water quality is discussed in Chapter 6.

6. The operating policy of the system has been limited to energy generation by constant-head reservoirs.

7. Alternative sources of water and alternative ways of satisfying water needs can be included in the model if they exist, but have not been specifically included. In particular, ground water, used solely for crop irrigation (as is the current practice in the high plains of West Texas) or in conjunction with surface water supplies, is an example of an alternative source. See the work of Milligan [1970] for typical relations to use in the optimization of the conjunctive use of ground water and surface water. Other ways of meeting water requirements are ground-water recharge, recycling of industrial water, reclamation of waste water, and avoidance of flood damage by flood-plain zoning.

The water resources system itself will be assumed to consist of a river

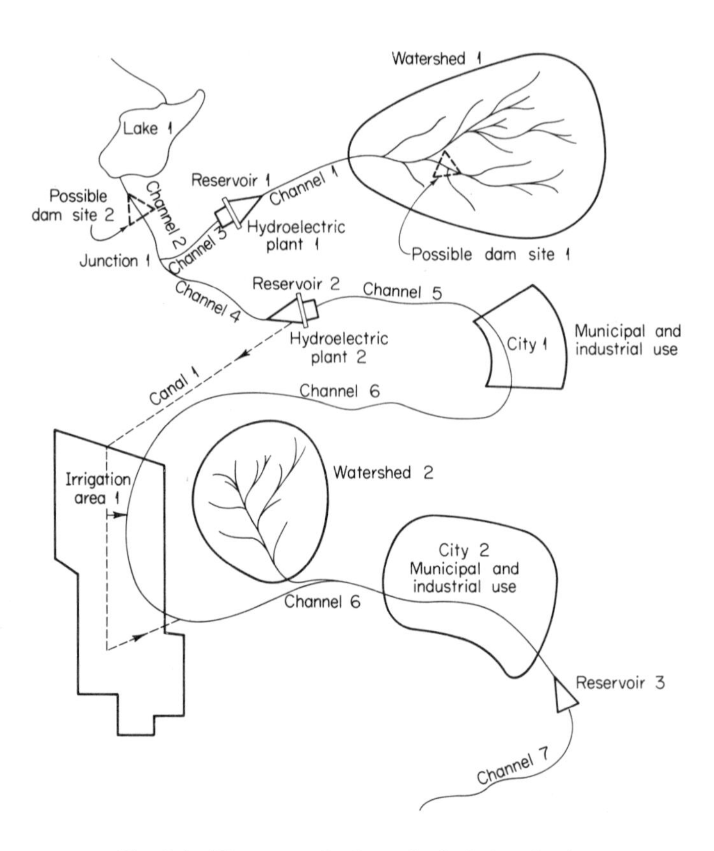

Fig. 2.1 Elements of a hypothetical river basin.

basin as illustrated in Fig. 2.1. Each subsystem of the river basin has definite demands on and consumptions of water. A forecast of increasing demand for municipal and industrial use, irrigation, energy, and recreation is available over a planning horizon of $T_{\max}$ years. Because the existing river (canal) and flow regulation facilities are operating at near optimal levels, it is presumed that these facilities will not be adequate to meet future demands.

A number of possible dam sites are available for the further regulation of river (canal) flows in the basin and/or the regulation of waters imported into the basin. Each dam can be built to several scales; each scale is considered a separate project. With each dam is associated a capital cost and an annual return associated with its optimal maximum use. The planner is concerned with what projects to undertake and when to undertake them. A capital budgetary constraint exists.

Therefore, the problem becomes: Given a planning horizon $T_{\max}$ and a set of alternative projects, select a period, if any, when each project will be introduced so that the objective function will be optimized while (1) staying within the budget limit, (2) meeting institutional constraints, (3) meeting all demands, and (4) satisfying all physical constraints.

2.2. Formulation of the Objective Function

The criteria that are established and used for project justification will be influenced by the instutional processes through which funds for development are authorized. In the United States, at the federal level, the agencies in charge of water resources development present each project to the Bureau of the Budget and to the Congress for individual authorization and approval. The main reason for carrying out an economic evaluation is to show that the project will produce at least as much in benefits as it will cost. Section 1.3 has discussed many of the criteria that must be considered in formulating the objective function. We will select as the criterion the maximization over the set of alternative projects of the sum of the discounted present value of net earnings. The objective function thus takes cognizance of the time value of money [Lesso, 1967].

In words the objective function comprises the difference between two classes of net returns (benefits) and two classes of costs:

$$f = \text{net operating returns} \begin{Bmatrix} \text{(1) from present system} \\ \text{(2) from additions} \end{Bmatrix} - \text{costs} \begin{Bmatrix} \text{(1) capital costs} \\ \text{(2) imported water costs} \end{Bmatrix}$$

In symbols we want to maximize the objective function

$$f = \sum_{t=1}^{T_{\max}} \alpha \sum_{j=1}^{N} \sum_{i=1}^{12} X_{ijt} + \sum_{t=1}^{T_{\max}} \alpha \sum_{j=N+1}^{M} \beta_{jt} \sum_{i=1}^{12} X_{ijt}$$

the *net operating return* from the present set of subsystems over the planning period; the *net operating return* from the newly added subsystems

$$- \sum_{t=1}^{T_{\max}} \alpha \sum_{j=N+1}^{M} \lambda_{jt} C_{jt} - \sum_{t=1}^{T_{\max}} \alpha \sum_{j=1}^{M} \sigma_{jt} K_{jt} \tag{2.1}$$

the *capital* cost of projects over the planning period; the *capital cost of providing* canals for imported water over the planning period where

C_{jt} = capital needed for building reservoir j in year t
K_{jt} = capital needed for building canal j in year t
M = maximum number of dams that can be built
N = number of dams that exist at the beginning of the planning horizon
$T_{\max}$ = length of the planning period
X_{ijt} = return from reservoir j in month i of year t
α = discount factor
β_{jt} = a Heavyside function; 1 designates that a return is available from project j in year t, while 0 indicates that no return is available
λ_{jt} = a Heavyside function; 1 designates capital must be provided to build dam j in year t, while 0 indicates capital does not have to be provided
σ_{jt} = a Heavyside function; 1 designates that capital must be provided to build a canal to supply imported water to reservoir j in year t, while 0 indicates that capital does not have to be provided

The subscript i refers to the month of operation, the subscript j to a particular dam, and the subscript t to the year of operation. The units for each of the symbols will be found in the list of notation in Appendix C. Other benefits and costs can be added to the objective function by analogy with the given terms. A specific example of an objective function will be found in Section 4.3.

2.3. Constraints

Constraints exist that limit the range of variation of each of the variables, prescribe their relationships to each other, and delineate the external influences on the planning. Constraints generally are of two basic types: equality or inequality constraints. Another type of constraint, as we shall see, is the restriction of a variable to being either 0 or 1. An exact mathematical representation of a water resources development project, even if possible, would lead to hopeless mathematical complexity. Therefore, in writing down the constraints it is necessary to attain a reasonable balance between accurate representation and mathematical manageability. Some variables have been deliberately omitted from the constraints because their impact on the optimum design is small while their contribution to the mathematical complexity is large. Other variables that are continuous, such as the river flow, have to be treated as discrete. However, it is believed that the model developed here is a reasonably accurate representation of a multipurpose water resources system and contains the variables that are the most relevant for optimal planning.

2.3.1. *Budgetary Constraints*

The capital budgetary constraint is calculated differently in the private and public sectors. In the private sector it is considered to be a function of a corporation's current assets and current debt level. In the public sector it is dependent upon congressional or state water resources appropriations. While it is clear that constraints on capital spending exist, the quantitative formulation of such constraints is quite subjective. Here we will say that the budgetary constraint consists of an annual limit on the availability of capital for new construction from public or private sources:

$$\alpha \sum_{j=N+1}^{M} \lambda_{jt} C_{jt} \leq M_t \qquad \text{for all } t \tag{2.2}$$

In essence we have said that in any year the appropriated funds for capital investment will not exceed M_t dollars.

2.3.2. *Institutional Constraints*

Instutitional constraints limit the number of dams that can be built at any site or in any year. We will assume somewhat arbitrarily that (1) only one new reservoir may be built in any year, and (2) each reservoir may be built in only one of the years. The mathematical statements for the institutional constraints are

$$\sum_{j=N+1}^{M} \lambda_{jt} \leq 1 \qquad \text{for all } t \tag{2.3}$$

i.e., at most only one new dam is built in any year, and

$$\sum_{t=1}^{T_{\max}} \sum_{j=N+7}^{N+9} \lambda_{jt} \leq 1 \tag{2.4}$$

Constraint (2.4) is an example of an inequality that excludes those combinations of projects that are technically infeasible. It differs from inequality (2.3) because it states that only one of the three projects ($N + 7$,

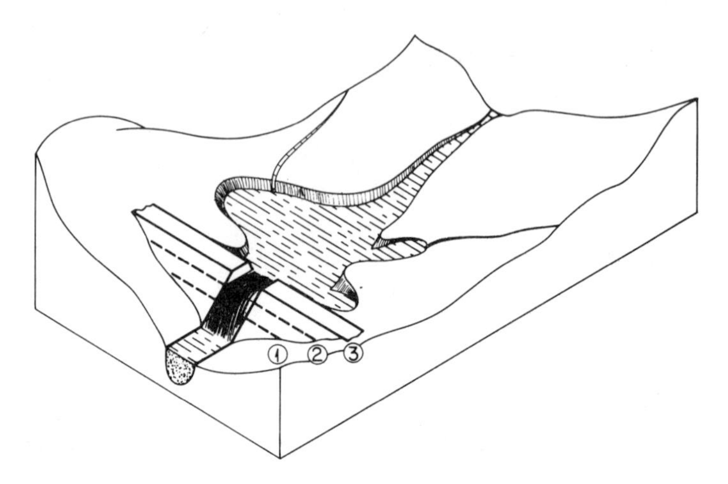

Fig. 2.2 Each project may represent one dam or a different sized dam at one site.

$N + 8$, $N + 9$) may *ever* be built. Figure 2.2 shows how these three projects may represent, for example, three different sizes of a dam at a specified dam site. In addition

$$\sum_{t=1}^{T_{\max}} \lambda_{jt} \leq 1 \qquad \text{for all} \quad j = N+1, \ldots, M \tag{2.5}$$

i.e., project j can be built in only one of the years, if it is built at all. Also,

$$\lambda_{jt} = 0 \text{ or } 1 \tag{2.6}$$

as defined in Section 2.2.

2.3.3. Dam Revenue Equation

The revenue for the operation of each reservoir can be represented by

$$X_{ijt} = f_j(S_{i+1,jt}, I_{ijt}, \sum_m A_{jm}Q_{imt}, D_{ijt}) \qquad \text{for all } j \tag{2.7}$$

where D_{ijt} is the quantity of water supplied for irrigation from reservoir j in month i of year t, f_j the return function for reservoir j, I_{ijt} the amount of imported water supplied to reservoir j in month i of year t, Q_{imt} the flow in stream or canal m during month i of year t, and S_{ijt} the storage volume of water in reservoir j at the beginning of month i of year t, i.e., the carry-over storage. The form of equation can be the same for each reservoir, but the coefficients will differ. The simplest form of the equation would be a linear sum of the revenues less the costs; a nonlinear equation introduces considerably more trouble in the solution. An example of the linear form of Eq. (2.7) that includes revenue from irrigation and energy sales less the operating cost of imported water is, for reservoir j,

$$X_{ijt} = \underbrace{\delta_{ijt}D_{ijt}}_{\text{net revenue from irrigation}} + \underbrace{\omega_{ijt}K_j \sum_m A_{jm}Q_{imt}}_{\text{net revenue from energy sales}} - \underbrace{\lambda_{ijt}H_{ijt}I_{ijt}}_{\text{net cost of importing water}}$$

where

A_{jm} = 1 if flow in link m enters reservoir j
= −1 if flow in link m leaves reservoir j
= 0 otherwise

H_{ijt} = cost coefficient, i.e., the operating cost of supplying imported water to reservoir j in month i of year t

K_j = amount of energy produced by turbine j per acre-foot of water

S_{ijt} = net revenue coefficient for the irrigation water supplied using a predetermined crop mix by reservoir j in month i of year t

δ_{ijt} = net revenue coefficient for irrigation water supplied by reservoir j in month i of year t

λ_{ijt} = 1 if water is imported to reservoir j in month i of year t, or = 0, otherwise

ω_{ijt} = net revenue coefficient for the energy generated by reservoir j in month i of year t

Because recreation benefits are not quantified and also because of the assumed linear relationship between energy and net water outflow from the reservoir, the linear dam revenue function X_{ijt} does not depend on the volume of water in the dam at the beginning of period $(i + 1)$, $S_{i+1,jt}$.

The total annual net revenue function for all the reservoirs in the system $(\hat{X}_t)$ is found by summing X_{ijt} over i and j. For the linear function one obtains

$$\hat{X}_t = \sum_{i=1}^{12} \sum_{j=1}^{M} X_{ijt}$$

$$= \sum_{i=1}^{12} \sum_{j=1}^{M} \delta_{ijt}D_{ijt} + \sum_{i=1}^{12} \omega_{ijt} \sum_{j=1}^{M} K_j \sum_{m=1}^{M_1} A_{jm}Q_{imt} - \sum_{i=1}^{12} \sum_{j=1}^{M} \lambda_{ijt}H_{ijt}I_{ijt}$$

Reservoirs are constructed on natural stream channels in order to provide some kind of regulation of the flow rate in those channels. The construction and operation of reservoirs essentially serve two purposes: first, the retention of upstream flow, and therefore, second, the regulation of downstream flow; or stated another way, the storage of excess water, and the later release of that water for beneficial uses. Thus reservoirs are generally classified according to whether they were designed for a single purpose or for multiple purposes; single-purpose reservoirs are usually simpler in design and operation than are multipurpose ones.

In those reservoirs built with flood control as a major purpose, it is essential that the reservoir capacity reserved for storage of flood water be emptied as soon as practical after the flood. In some cases, because of a definite seasonal pattern of floods, stored floodwater may also be retained, at least to the extent possible, for later conservation uses. In those reservoirs built for conservation uses, stream flow in excess of current requirements is stored in the reservoir and not released until later when it is needed. In this manner reservoirs serve the multiple purpose of flood prevention in times of high flows and supply of necessary water in times of low flows. The release of stored water may be used for many purposes, such as energy generation, irrigation, municipal water supply, and others. All of these uses are generally referred to as conservation uses, since the excess water is conserved, or stored, for them.

Consider first the benefits from irrigation water. If the willingness to pay some contract price, or the market price, is the determining factor for the value of irrigation water, then D_{ijt} multiplied by the price is the value of the irrigation water from reservoir j in month i of year t. On the other hand, if the "with or without" principle is to be used in determining the revenue from irrigation water, as discussed in Section 1.3, the benefits are calculated by comparing the situation without the project to the conditions that will prevail with the irrigation water.

In such an analysis farms having similar soil, size, and transportation costs are grouped together. For each homogeneous group of farms, average production coefficients, costs, and returns for various potential agricultural activities are estimated. Based upon information from homogeneous farm groups and availability of resources, relations such as the following can be prepared for each group both before and after the introduction of irrigation water:

$$R_{\text{net}} = \sum_{k=1} \sum_{l=1} p_{kl} y_{kl} q_{kl} - \sum_{k=1} \sum_{l=1} c_{kl} y_{kl} q_{kl}$$

where

R_{net} = net revenue per unit volume of water from the irrigation water
c_{kl} = cost per ton of crop k by method l
p_{kl} = market value per ton of crop k by method l
q_{kl} = net consumptive use of irrigation water (area per unit volume) in each growing season for crop k by method l
y_{kl} = production of crop k by method l

For example, if the crop is sugar beets and the following data are established for one class of farms for 0.24 acre-ft of irrigation water per acre of crop land

Land required: 0.09 acre/ton = $(1/y_{11})$
Costs, inclusive of labor and land: \$2.37/ton
Market price: \$28.50/ton

then the net revenue from the farm before new irrigation water was available would be (omitting a charge for the land)

$$(1/0.24)[\$28.50 - \$2.37](1/0.09)$$

By adding more water the yield y_{11} would be increased; hence the maximum value of the water could be estimated, given the water versus yield relationship. If the latter were linear, Eq. (2.7) would have a linear term involving D_{ijt}; but if the relationship were nonlinear, then Eq. (2.7) would be nonlinear in D_{ijt}.

A second contributor to revenue from the operation of a reservoir is the generation of hydroelectric power. The amount of electricity generated by a hydroelectric plant associated with a reservoir depends upon the installed capacity of the plant, the inflow of water into reservoir, the available storage capacity in the reservoir, the level of the water in the reservoir, and the mandatory releases to meet other downstream requirements for water. Installed capacity may only indicate the potential or theoretical value for power generation. It is necessary to distinguish between "dependable" or "firm" power, which is the continuous output capacity available throughout every year, and "secondary" power, which is power available intermittently or for only portions of the year because of the inadequacy of inflow and water-storage capacity, or because of large downstream flow requirements for irrigation, pollution control, navigation, and so on. Therefore, it is customary for the value of electricity from a hydro plant to be expressed in terms of two components: a value in mills per kilowatt hour for "firm" energy and another value for "secondary" energy.

In any power system there are changes in demand for electricity over any 24 hours. The ratio of peak demand to average demand, which is called the system-load factor, is generally a function of the composition of the customers in a system. For example, the larger the relative magnitude of the electricity consumed by industrial consumers to householders the smaller the system-load factor will be. It might seem logical to use plants having the lowest operating costs to provide the base load, and plants with higher operating cost to meet energy requirements at peak load. Since operating costs of hydro plants are virtually zero, it would be ex-

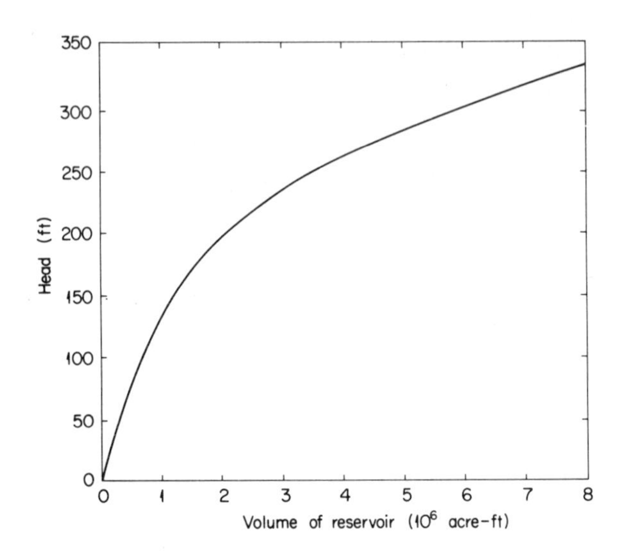

Fig. 2.3 Relation between storage volume of a reservoir and head.

pected that in mixed hydro–thermal systems the hydro plants normally would be assigned to base-load duty. However, in practice hydro plants are more likely to be used to meet peaking requirements rather than for the base load because (1) limitations on the available storage capacity of the reservoir and large mandatory releases and other requirements are such that the hydroelectric plant is not capable of providing the base load, and (2) the time required to start or shut down the generation of electricity in a hydroelectric plant is relatively short compared to that for a thermal plant. It is more economical to start and shut down a hydroelectric plant than a thermal plant.

The net revenue for a period from electric energy might be expressed as follows (in words):

$$\text{net revenue} = (\text{dollars/MW-hr})(\text{energy output}) - \text{generating costs}$$

and the energy output might be

$$\text{energy output} = 8.50\,\gamma\Big(\sum_m A_{jm}Q_{imt}\Big)Z_{ijt}$$

where Z_{ijt} is the head in reservoir j in month i of year t and 8.50 is a constant appropriate for the reservoir. If some of the water released by-passes the turbines, then the fraction passing through the turbines γ is less than 1.

The head Z_{ijt} can be related to the storage in the reservoir by means of empirical data for each reservoir, such as shown in Fig. 2.3. The figure represents a least-squares fit of typical empirical data.

$$Z_{ijt} = 9.5367S_{ijt} - 15.180S_{ijt}^2 + 1.1311S_{ijt}^3 - 0.0297885S_{ijt}^4$$

On the other hand, if the reservoir is a constant-head reservoir, the energy output is then directly proportional to the reservoir outflow. Maass. [1962] indicated that the operating costs of power plants are essentially proportional to the size of the plant except at very low power capacities. We will assume that the sizes of the power plants are given; hence they will not be variables.

2.3.4. Recreation Demands and Related Constraints

Some of the major benefits to be derived from the expansion of a water resources system are the primary and secondary benefits to be derived from recreation. Therefore, it appears important that models developed for the analysis of such systems contain factors that represent the demand for and supplies of recreation in a river basin. Recreational benefits can be related to five factors: (1) the accessibility of recreational resources to the public, (2) the relative attractiveness of the resources, (3) the competing opportunities available, (4) the capacity of the facilities and resources to accommodate the public, and (5) the willingness of the public to incur expenses to enjoy the use of the recreational resources, if they exist. How to measure these quantities and introduce them into a model of a river basin is not entirely resolved, as one might expect. Some inroads are being made [Ditton, 1969], but it is safe to say that we do not have enough current knowledge to quantify the recreation factors to the extent desired. Therefore, we will assume that the following constraints are appropriate in placing limits on water storage for recreation in reservoirs, but we will not introduce terms into the objective function representing recreation benefits. However, if recreation benefits *must* be included in the objective function, one might take data for attendance versus reservoir capacity [U.S. Corps of Engineers, 1962], fit it by regression

$$\frac{\text{no. attendees}}{\text{surface acre}} = 3.65 - 0.31S_j + 0.012S_j^2$$

evaluate each visitor day, say \$1.00/day, and, using the relation between surface acreage and storage such as Fig. 2.4 plus estimated attendance, estimate the recreation benefits per month (or year).

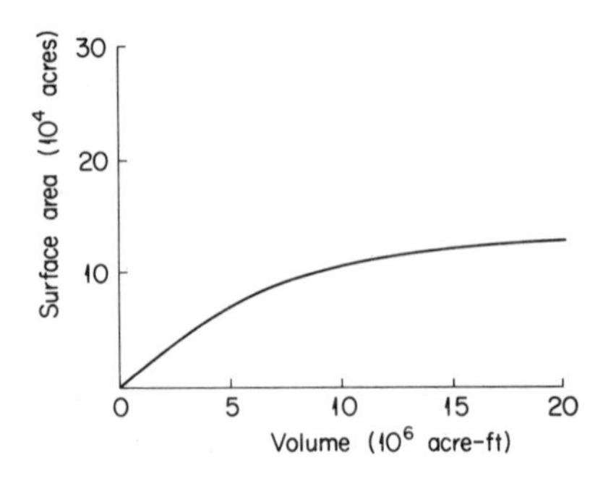

Fig. 2.4 Relation between surface area of a reservoir and storage volume S_j. Data are based on a topographical map prepared prior to the filling of the reservoir.

We will assume that the storage volume of water in reservoir j at the beginning of month i of year t, S_{ijt}, must be equal to or greater than some specified value:

$$S_{ijt} \geq \tilde{S}_{jr} \qquad \text{for all } i, j, \text{ and } t \tag{2.8}$$

where $\tilde{S}_{jr}$ is the minimum acceptable recreation level for reservoir j. When the water volume drops below $\tilde{S}_{jr}$, unsightly mud-flats appear and so the reservoir becomes unsuitable for recreation. In addition, because of the interconnection between the recreational sites in the river basin,

$$\sum_{j=1}^{4} f_j(S_{ijt}) + \sum_{j=N+1}^{N+3} f_j(S_{ijt}) \geq R_{it} \qquad \text{for all } i \text{ and } t \tag{2.9}$$

where R_{it} is the minimum acceptable recreation needs for a certain section of the river basin and is a function of the storage levels $[f_j(S_{ijt})]$ of the reservoirs in that section. In Eq. (2.9) it is assumed that four reservoirs already exist and that three new dams, $N + 1$, $N + 2$, and $N + 3$, may be added.

2.3.5. Municipal and Industrial Demands and Related Constraints

Municipal and industrial demands for water lead to a sequence of needs that must be met over the planning horizon. Municipal constraints are *mandatory* because their true quantitative worth has not been found. The fulfillment of industrial, irrigation, or energy demands is not so critical, since a penalty can be incorporated in the objective function for those occasions when the demands are not met. For the energy demand we have the constraint

$$\sum_j f_j(S_{i+1,jt}, \sum_m A_{jm}Q_{imt}) \geq P_{it} \qquad \text{for all } i \text{ and } t \tag{2.10}$$

where

f_j = energy generation function for reservoir j
P_{it} = minimum acceptable energy demand for the *total* river basin (which will increase over the years)
A_{jm} = 1 if flow in link m enters reservoir j
= -1 if the flow in link m leaves reservoir j
= 0 otherwise
Q_{imt} = flow in link m during month i of year t

For the example problem in Chapter 4, the specific function used is listed in Section 3.4.

Future demands for irrigation, recreation, and municipal and industrial use of water can be met only by reservoirs in the locality adjacent to each municipality, industry, or irrigation area. Consequently, the river basin is divided into a number of subsections for these purposes, and the irrigation, municipal, industrial, and recreation demands for each subsection may be met only by dams in that subsection. The demands will show an increasing value over the years.

Constraints (2.11) and (2.12) are inequalities that deal with municipal, industrial, and irrigation demands of an arbitrary subsection. It is assumed that four reservoirs already exist in the chosen subsection and three new dams ($N + 1$, $N + 2$, $N + 3$) may be added. The same constraints apply to *every* subsection except that summation is over different reservoirs:

$$\sum_{j=1}^{4} F_{ijt} + \sum_{j=N+1}^{j=N+3} F_{ijt} \geq \tilde{F}_{it} \qquad \text{for all } i \text{ and } t \tag{2.11}$$

where $\tilde{F}_{it}$ is the minimum municipal and industrial demand for a subsection of the river basin, and F_{ijt} is the water supplied for municipal and industrial use from reservoir j in month i of year t. Also

$$\sum_{j=1}^{4} D_{ijt} + \sum_{j=N+1}^{N+3} D_{ijt} \geq G_{it} \qquad \text{for all } i \text{ and } t \tag{2.12}$$

where G_{it} is the minimum irrigation need for a subsection of the river basin.

2.3.6. *Physical Constraints*

Typical physical constraints that must be introduced into the model of the river basin are (1) bounds on river (canal) flows, and (2) mass balances on each reservoir. These physical constraints limit the flow of water through the system so as to satisfy the conservation of mass balances on the quantity of water as well as to incorporate the physical capacities of the arcs (rivers, canals) in the network. Various degrees of detail can be incorporated in the physical constraints, but the mass balances may include terms representing (1) runoff–rainfall in the watersheds, (2) channel flow at downstream points for an upstream input, (3) groundwater models, (4) water control works, such as reservoirs or lakes, including evaporation, (5) factors for municipal and industrial net uses and crop irrigation, and (6) requirements for such factors as navigation, recreation, and water quality.

While the detail of the river basin mass balances can be as complex or as simple as the analyst desires, obviously there exists some optimum degree of complexity in which a compromise is found between the accuracy of the model representation and the ease of its solution. The model developed here is a compromise. It encompasses sufficient subsystems and elements to be easily extended to represent a real hydrological system, yet its solution is not so complicated as to make its use impractical.

A hypothetical river basin rather than an actual one was selected for the modeling in order to make the model as general as possible. No actual basin contains precisely the elements incorporated in the model, although many of the elements are found in all basins. By forming a hypothetical system containing elements from many different river basins, we obtain a more flexible model than would be possible using a single, real river basin.

Figure 2.1 shows a hypothetical river basin that includes the following major features: (1) channel streamflow, (2) canal flow, (3) runoff–rainfall input, (4) reservoirs (existing and potential), (5) stream junctions, (6) a constant-level lake, (7) irrigation removal, (8) municipal removal and return, and (9) industrial removal and return. Figure 2.1 does not include the following features that may be present in some water resources systems: (1) a bay or estuary, (2) groundwater supply and/or recharge, (3) evaporation and transpiration (except as included in a specific subsystem), (4) sedimentation, (5) navigational requirements, (6) flood control requirements, (7) water quality requirements, or (8) recreational and wildlife requirements.

Because all the basin subsystems are connected by the flow of the river downstream, there are no feedback or recycle loops of information, and it is easy to connect the subsystem elements to form the total basin model. As a practical matter, it is important for all the elements to have common dimensions (units) and time bases in the flow of information connecting the subsystem models through common variables. For example, the release from a reservoir, which is the output for a reservoir model, becomes an input variable to the adjacent downstream channel flow model. Obviously,

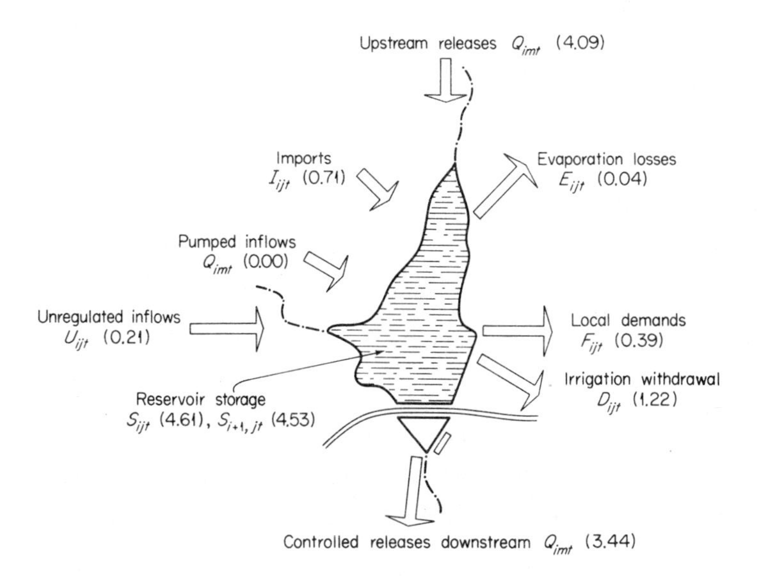

Fig. 2.5 Monthly material balance on a reservoir. The numbers in parentheses give typical monthly flows and volumes in hundreds of thousands of acre-feet.

the reservoir output must be equal to and consistent with the channel flow input.

Equation (2.13) is the monthly mass balance for each reservoir. Figure 2.5 illustrates the terms involved in the balance. The first term on the right-hand side is the sum of all regulated inflows and outflows to and from the reservoir other than those specified by the other terms.

$$\underbrace{S_{i+1,jt} - S_{ijt}}_{\text{accumulation}} = \underbrace{\sum_{m=1}^{M_1} A_{jm} Q_{imt}}_{\text{net regulated flow}} - \underbrace{F_{ijt}}_{\text{municipal and industrial use}} - \underbrace{E_{ijt}}_{\text{evaporation losses}} - \underbrace{D_{ijt}}_{\text{irrigation withdrawal}} + \underbrace{U_{ijt}}_{\text{unregulated flow}} + \underbrace{\lambda_{ijt} I_{ijt}}_{\text{imported water}} \qquad (2.13)$$

Any junction of two or more flows would reduce to an equation of the following form:

$$\sum_{m=1}^{M_1} A_{rm}Q_{imt} - U_{irt} = 0 \tag{2.13a}$$

where the subscript r represents all the flows entering and leaving junction r. Examples of Eq. (2.13) can be found in Section 1.3; note that the equation is linear.

Evaporation losses from the surface of a reservoir can be a significant portion of the total inflow, and consequently evaporation losses must be included in the mathematical model. To calculate evaporation losses, one must determine the evaporation coefficients, which are the average monthly evaporation rates for various months of the year measured in feet of water per time period. Because the net evaporation loss from the reservoir surface is an empirical quantity dependent on climatic conditions such as weather, geographical location, and time of year, the evaporation coefficient is related to temperature, precipitation, humidity, and wind movement. Kane [1967], for example, has tabulated common values of the evaporation coefficient, as have Lowery [1960], Mobasheri [1968], and Hall and Dracup [1970]. For example, for the area of Texas in which the Colorado River Highland Lakes are located, the coefficient has a value of approximately 0.42 ft/month averaged over the entire year, but has a value of 0.55 in the summer months of May–October, and a value of 0.29 for the rest of the year. For the summer months the evaporation loss for reservoir j might be

$$E_{ijt}\,\frac{\text{acre-ft}}{\text{month}} = \left(\frac{0.55\text{ ft}}{\text{month}}\right)\left(\begin{matrix}\text{area of reservoir } j\\ \text{in acres}\end{matrix}\right)$$

Figure 2.4 shows how the area of reservoir j can be related to the storage S_{ijt}. Reservoirs and lakes connected to the ground-water table and with subsurface inlets and outlets need to have a much more complex term for E_{ijt}, but the quantities involved are very difficult to measure or estimate, and hence have been omitted here.

As to the other variables in Eq. (2.13), F_{ijt}, D_{ijt}, U_{ijt}, and I_{ijt} must be specified by schedules of supply and demand. Imports of water are presumed to be limited in quantity:

$$\sum_j I_{ijt} \le J_{it} \qquad \text{for all } i \text{ and } t \tag{2.14}$$

where J_{it} is the maximum available imported water for month i of year t.

Associated with each reservoir is an initial storage volume

$$S_{1j1} = \tilde{S}_{j0} \qquad \text{for all } j \tag{2.15}$$

where $\tilde{S}_{j0}$ is the specified initial volume (initial condition) for reservoir j.

Finally, upper and lower bounds on the variables Q_{imt} and S_{ijt} must be specified because the river and canal flow rates are limited, as is the reservoir storage, as explained in Section 1.3:

$$L_m \leq Q_{imt} \leq C_m \tag{2.16}$$

$$0 \leq S_{ijt} \leq V_j \tag{2.17}$$

2.4. Summary of the Objective Function and Constraints

The purpose of this section is to summarize the objective function and constraints that comprise the model of the river basin system.

Objective Function

Maximize

$$\begin{aligned} f = &\sum_{t=1}^{T_{\max}} \alpha \sum_{j=1}^{N} \sum_{i=1}^{12} X_{ijt} + \sum_{t=1}^{T_{\max}} \alpha \sum_{j=N+1}^{M} \beta_{jt} \sum_{i=1}^{12} X_{ijt} \\ &- \sum_{t=1}^{T_{\max}} \alpha \sum_{j=N+1}^{M} \lambda_{jt} C_{jt} - \sum_{t=1}^{T_{\max}} \alpha \sum_{j=1}^{M} \sum_{i=1}^{12} \sigma_{jt} K_{jt} \end{aligned} \tag{2.1}$$

Budgetary Constraints

$$\alpha \sum_{j=N+1}^{M} \lambda_{jt} C_{jt} \leq M_t \qquad \text{for all } t \tag{2.2}$$

Institutional Constraints

$$\sum_{j=N+1}^{M} \lambda_{jt} \leq 1 \qquad \text{for all } t \tag{2.3}$$

$$\sum_{t=1}^{T_{\max}} \sum_{j=N+7}^{N+9} \lambda_{jt} \leq 1 \tag{2.4}$$

$$\sum_{t=1}^{T_{\max}} \lambda_{jt} \leq 1 \qquad \text{for all } j = N+1, \ldots, M \tag{2.5}$$

$$\lambda_{jt} = 0 \text{ or } 1 \tag{2.6}$$

Dam Revenue Equation

$$X_{ijt} = f_j(S_{i+1,jt}, I_{ijt}, \sum_j A_{jm}Q_{imt}, D_{ijt}) \qquad \text{for all } j \tag{2.7}$$

Recreation Constraints

$$S_{ijt} \geq \tilde{S}_{jr} \qquad \text{for all } i, j, \text{ and } t \tag{2.8}$$

$$\sum_{j=1}^{4} f_j(S_{ijt}) + \sum_{j=N+1}^{N+3} f_j(S_{ijt}) \geq R_{it} \qquad \text{for all } i \text{ and } t \tag{2.9}$$

Municipal and Industrial Requirements

$$\sum_j f_j(S_{i+1,jt}, \sum_m A_{jm}Q_{imt}) \geq P_{it} \qquad \text{for all } i \text{ and } t \tag{2.10}$$

$$\sum_{j=1}^{4} F_{ijt} + \sum_{j=N+1}^{j=N+3} F_{ijt} \geq \tilde{F}_{it} \qquad \text{for all } i \text{ and } t \tag{2.11}$$

$$\sum_{j=1}^{4} D_{ijt} + \sum_{j=N+1}^{N+3} D_{ijt} \geq G_{it} \qquad \text{for all } i \text{ and } t \tag{2.12}$$

Physical Constraints

$$S_{i+1,jt} - S_{ijt} = \sum_{m=1}^{M_1} A_{jm}Q_{imt} - F_{ijt} - E_{ijt} - D_{ijt} + U_{ijt} + \lambda_{ijt}I_{ijt} \tag{2.13}$$

$$\sum_j I_{ijt} \leq J_{it} \qquad \text{for all } i \text{ and } t \tag{2.14}$$

$$S_{1j1} = \tilde{S}_{j0} \qquad \text{for all } j \tag{2.15}$$

$$L_m \leq Q_{imt} \leq C_m \tag{2.16}$$

$$0 \leq S_{ijt} \leq V_j \tag{2.17}$$

2.5. Sources of Data for the Mathematical Model

Table 2.1 lists in alphabetical order the constants and variables involved in the model of the water resources system together with the numbers of the equations and functions in which the variables appear. Table 2.2

Table 2.1

List of Constants in the Mathematical Model

Constant	Description	Equation in which it appears
α	Discount factor	(2.1), (2.2)
A_{jm}	Designates whether flow occurs	(2.10), (2.13)
C_{jt}	Capital to build reservoir	(2.1), (2.2)
C_m	Maximum flow capacity of river or canal	(2.16)
$\tilde{F}_{it}$	Minimum industrial and municipal demand	(2.11)
G_{it}	Minimum irrigation demand	(2.12)
J_{it}	Amount of imported water	(2.14)
K_{jt}	Capital to build canal	(2.1)
L_m	Minimum flow capacity in river or canal	(2.16)
M_t	Capital budget limit for year	(2.2)
M_1	Number of links in system	(2.13)
M	Maximum number of dams	(2.1)–(2.3), (2.14)
N	Initial number of dams	(2.1)–(2.3)
P_{it}	Minimum energy demand	(2.10)
R_{it}	Minimum acceptable recreational needs	(2.9)
$\tilde{S}_{j0}$	Initial reservoir volume	(2.15)
$\tilde{S}_{jr}$	Minimum reservoir volume for recreation	(2.8)
T_{max}	Length of planning period	(2.1), (2.4), (2.5)
U_{ijt}	Net unregulated flow to reservoir	(2.13)
V_j	Maximum capacity of reservoir	(2.17)

similarly lists the independent and dependent variables. In this section we point out the sources of information that can be used to obtain the values of the constants and known inputs to the model.

2.5.1. Runoff (Unregulated Flows)

Stream gauging in the United States is supervised by one federal agency, the U.S. Geological Survey (USGS), Department of the Interior. Gauging work is done under a long-standing 50–50 cooperative arrangement with other "sponsoring" agencies that pay half the cost of the gauging program. In Texas, for example, the gauging work is cosponsored by the Texas Water Development Board.

For convenience, the Geological Survey has divided the United States into its major river basins such as the Upper Mississippi Basin, the Ohio River Basin and the St. Lawrence River Basin. For each river basin the

Table 2.2

List of Variables in the Mathematical Model

Variable	Description	Equation in which it appears
Independent variables		
β_{jt}	Designated whether return is available	(2.1)
I_{ijt}	Amount of imported water	(2.7), (2.13), (2.14)
Q_{imt}	Quantity of water flow	(2.7), (2.10), (2.13), (2.16)
Dependent variables		
λ_{ijt}	Designates whether water is imported	(2.13)
λ_{jt}	Designates whether capital must be provided	(2.1), (2.2)–(2.6)
D_{ijt}	Water supplied for irrigation	(2.7), (2.12), (2.13)
E_{ijt}	Evaporation losses from reservoir	(2.13)
F_{ijt}	Amount of water supplied for industrial and municipal use	(2.11), (2.13)
S_{ijt}	Storage volume of water	(2.7)–(2.10), (2.13), (2.15), (2.17)
X_{ijt}	Revenue from reservoir	(2.1), (2.7)

Geological Survey has collected all the available flow data, prior to 1951, and published it in the form of compilation reports (e.g., U.S. Geological Survey [1957]). For years subsequent to 1951, the collected flow data is obtainable from the USGS annual Water Supply Papers for the same basins (e.g., U.S. Geological Survey [1961]).

In Latin America the available hydrological data are limited. A study of the relevant literature on countries as disparate as Colombia and Chile [Bulkley *et al.*, 1965; Wallace, 1966; Posada *et al.*, 1966; McLaughlin, 1967; King, 1967] reveals that the length of rainfall and streamflow records in *major* river basins can vary from 5 to 50 years.

In the model, streamflows from a typical historical hydrological record (U_{ijt}) can be used as inputs to constraint (2.13) as illustrated in Fig. 2.6.

2.5.2. *The Budgetary Constraint*

The budgetary constraint is of the form suggested by Marglin [1962] for participation by a national government in water resources projects.

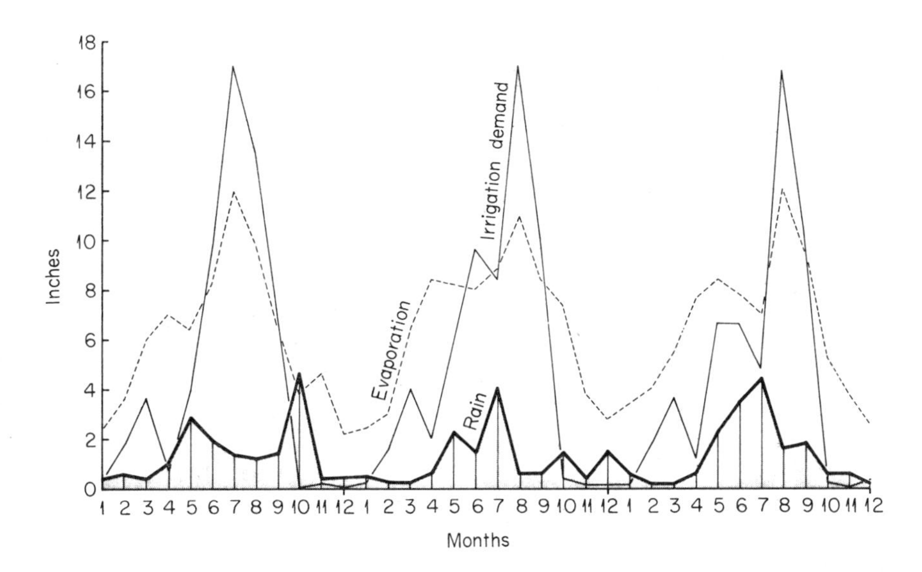

Fig. 2.6 Typical monthly data for runoff, evaporation, and irrigation. (From Texas Water Development Board [1971].)

This constraint is designed to limit the long-term flow of funds from the national treasury into water resource development. It represents the costs less the revenues and taxes stimulated by the project over its entire economic life.

In this book we assume that the government is willing to fund the construction by providing one lump sum at the beginning ($t = 0$) in the amount of $\hat{C}_0$. In any subsequent year t the present value of the funds available ($\hat{C}_t$) is

$$\hat{C}_t = \hat{C}_{t-1} + \hat{X}_{t-1} - \hat{I}_{t-1}/(1 + r)^{t-1} \tag{2.18}$$

where $\hat{X}_{t-1}$ is the revenue generated in year $(t - 1)$, $\hat{I}_{t-1}$ the investment in year $(t - 1)$, and r the interest rate. The model allows for different interest rates to apply to positive and negative debt levels. M_t in expression (2.18) is then equal to $\hat{C}_t$ or some fraction thereof. If the national government decides that funding be on an annual or some other basis, M_t can be easily modified to conform.

2.5.3. Water Use in the Various Economic Sectors

Water use data for the United States and its major regions have been given quinquennially since 1950 by the U.S. Geological Survey for the major water-using sectors and a number of specific industries (see U.S. Geological Survey Circ. [1951, 1956, 1961].

The United Nations, through its agencies, is one of the best sources of general information on water use. One of its agencies, the Economic Commission for Latin America (ECLA), has published a comprehensive series of reports on water use, sector by sector for each country [ECLA, 1961, 1964] as well as a general overview of the demand for water [ECLA, 1963a, b].

In the following paragraphs we will indicate appropriate sources of information for irrigation, municipal and industrial water use, hydroelectric energy generation, and the cost of recreation.

(a) IRRIGATION

For the United States, the most comprehensive source of irrigation data is the quinquennial U.S. Census of Agriculture [1969] that is published by the U.S. Department of Commerce, Bureau of the Census. Volume I provides data on irrigated acreage by crop type in each county. Volume III gives the total irrigated acreage within particular portions of a given river system. Abundant and useful information is also available from two other Federal agencies, namely the U.S. Department of the Interior, Bureau of Reclamation and the U.S. Department of Agriculture, Soil Conservation Service. For countries other than the United States the F od and Agricultural Yearbook [United Nations, 1970] collates irrigation, crop acreage, and production, nation by nation.

Irrigation revenue coefficients are needed in the model to calculate the return of Eq. (2.7); they are determined by finding from the historical records of the region of interest the net increase in benefits to that region when the change from dryland farming to irrigation farming was made (See Section 2.3.3).

Consumptive irrigation data are needed to calculate the profile of total irrigation demands [G_{it} in constraint (2.12)] such as shown in Fig. 2.6. Consumptive use is the total quantity of water per acre per time period (regardless of source) required by a given crop within some defined area for the full development of that crop. The actual water requirements are obtained by dividing the consumptive use by an efficiency factor that is the ratio of water consumed by the crop to the amount diverted into the irrigated area. Considerable research and experimentation has taken place to determine the consumptive use of water for crops. Agriculturists have spent much effort in attempting to relate the water requirements of crops to various atmospheric phenomena and soil properties, resulting in a variety of complex empirical methods. Some of the better-known methods for estimating the consumptive use include the methods of Blaney and Criddle [Criddle, 1958], Hargreaves [1957], and Penman [1956]. Of these, the Blaney–Criddle method is most widely used. To simplify the application of the model being used, the prediction of irrigation requirements in this book have not been determined from atmospheric and soil properties, but rather from tabulated values taken from the literature [McDaniels, 1960].

(b) MUNICIPAL AND INDUSTRIAL WATER USE

Municipal and industrial demand is directly used in calculating the profile of future municipal and irrigation demands [$\tilde{F}_{it}$ in constraint (2.11)]. The rate of water removal by a city is a function of such factors as population, economic characteristics of the population, general climate, degree of industrilaization of the city, and the local weather conditions. Warmer weather and drier weather tend to increase the rate of water removal, so that the higher rates come in the summer months for a particular city, as shown in Fig. 2.7. All municipal water supply agencies maintain detailed

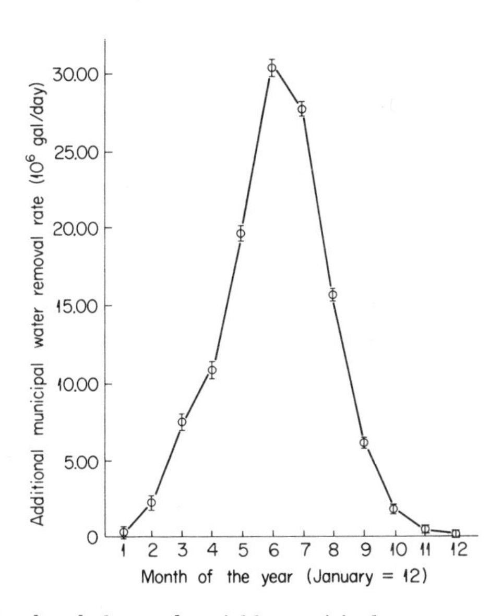

Fig. 2.7 Predicted and observed variable municipal water removal rate for Austin, Texas (monthly averages, 1961–1966). The base line of zero is at the level of the average January and February withdrawal rate. Solid lines—observed data; circles—predicted values; bars—range of 95% confidence limits.

records of water usage that can be related to such factors as the population, degree of industrialization, and general climate of the city and projected into the future [Howe and Linaweaver, 1967].

Agencies of international organizations such as the United Nations (UN) and the Organization of American States (OAS) have provided data on municipal [Pan American Health Organization/World Health Organization, 1963] and industrial water use [United Nations, 1958]. Additional data for Latin America have been published by the Center of Latin American Studies [1962].

Future industrial demand for water has been empirically evaluated by Olson [1966] and Sewell and Bower [1968]. In the United States, the quinquennial U.S. Census of Manufacturers, [1972] is the comprehensive source of past industrial water use data. Water used in the petroleum and mineral extraction sectors has also been estimated [Buttermore, 1966; Kaufman and Nadler, 1966]. Several other sources of primary data are available [State of California Department of Water Resources, 1964; National Association of Manufacturers and Chamber of Commerce of the United States, 1965; Texas Water Development Board, 1968]. Comprehensive data for industrial water use in California and Texas has been compiled [Lofting and McGauhey 1968].

(c) HYDROELECTRIC ENERGY

In the model the profile of future energy demands [P_{it} in constraint (2.10)] has been taken directly from the published data. Data for the generation of hydroelectric energy are a function of the characteristics of *each* individual dam: (1) the volume of dead-water storage behind the dam, (2) the land topography around the dam site, and (3) whether or not the water behind the dam is kept at a constant level. The energy revenue coefficients that are used for the calculation of the return in Eq. (2.7) are the wholesale unit prices quoted by local agencies currently selling hydroelectric energy. Published data on hydroelectric energy have concentrated on two factors: (1) the maximum availability of hydroelectric energy in a given area, and (2) the present rate of increase of availability of hydroelectric energy (e.g., United Nations [1962], ECLA [1967]). In the United States, the principal sources of data on the hydroelectric energy generation characteristics of dams are publications of the U.S. Department of Defense, the U.S. Army Corps of Engineers, and the U.S. Department of the Interior, Bureau of Reclamation.

(d) RECREATION

Recreation benefits have been calculated in terms of the projected number of persons coming to a reservoir over a period of time [Clawson and Knetsch, 1966; Merewitz, 1966; Tussey, 1967; Sirles, 1968; Stewart and Fraser, 1969]. These benefits have been correlated with the surface area of the reservoir and the month of the year. By this analysis the profile of future recreation demands [R_{it} in constraint (2.9)] and the minimum recreation level for each reservoir [$\tilde{S}_{jr}$ in constraint (2.8)] can be determined. A profile of future recreational visits to the reservoirs in the area was assumed, and the surface reservoir area needed was calculated from the correlations available for the reservoirs.

2.5.4. Costs of Reservoirs

For any selected reservoir design the capital cost of the reservoir is essentially the same, no matter what the purpose of the dam. Costs include land acquisition and clearing, relocation of structures and roads, the dam itself and its allied works, and diking, access, and service facilities. Design and estimating procedures for dams and reservoirs are well established, and rapid improvements in estimation techniques are being made as computer technology is utilized increasingly for this task (e.g., Weaver [1963]). Costs for actual dams in Scotland (Clatteringshaws Reservoir) and in the Clearwater and Lehigh River Basin in the United States have been reported by Guthrie [1958], Bower [1962], and Hufschmidt and Fiering [1966]. In each case the cost of the reservoir was correlated with the capacity of the water behind the dam.

The capital costs of new projects [C_{jt} in the objective function (2.1)] and also the capacities of these reservoirs [V_j in constraint (2.17)] have been taken directly from the data cited above.

2.5.5. Importation of Water

The model has incorporated a term to allow for the importation of water from without the basin to meet in-basin needs. This water could be transferred on an interbasin, interstate, or international basis. The unit cost of imported water [H_{ijt} in the cited example of Eq. (2.7)] and maximum allowable amount of imported water [J_{it} in constraint (2.14)] are arranged by compact or treaty between the entities involved. In the example solution of the optimal expansion problem described in Chapter 4, "reasonable" values of H_{ijt} and J_{it} are assumed.

2.5.6. Evaporation Losses

Evaporation losses from the surface of a reservoir can be up to 10% of the total inflow, so that these losses have been included in the mathematical model [variable E_{ijt} in Eq. (2.13)]. The procedure used to calculate the evaporation losses has been described in Section 2.3.6.

2.5.7. Other Constants

(a) DISCOUNT FACTOR

The discount factor [variable α in the objective function (2.1) and in inequality (2.2)] depends on the discount rate r

$$\alpha = [1/(1 + r)]^t$$

In the United States this rate is determined by the Water Resources Council, and in 1970 was 4.625%.

(b) BOUNDS ON ARC FLOWS

The capacity of an arc [C_m in constraint (2.16)] is the maximum amount of water that can flow through the arc (a reach of river or a canal) without flooding. The lower bound on arc flow (L_m) is usually zero, but many have a nonzero value when water quality standards or navigation requirements demand a definite flow of water in that arc. In the example discussed in Chapter 4, all of the values of C_m and L_m are assumed.

(c) INITIAL RESERVOIR CONDITIONS

The initial condition for each reservoir [$\tilde{S}_{j0}$ in equation (2.15)] have been assumed.

(d) OTHER CONSTANTS

Values for the number of reservoirs initially present in the system (N in the objective function and in various summations), the maximum possible number of reservoirs in the systems (M in the objective function and in various summations), the total number of links in the system [M_1 in constraint (2.13)], and the length of the time horizon ($T_{\max}$ in the objective function) are assumed in the example problem.

2.6. Summary

In this chapter we have examined the preparation of a model of a water resources system and examined the underlying assumptions that govern the performance of the model. An appropriate time scale (the month) was selected for the water flows, and a time scale of a year was selected for the introduction of new projects. All the variables and parameters were restricted to be deterministic in order to simplify the model and to make it possible to apply the optimization techniques described in Chapter 3.

The important features of the model that make a solution possible in a reasonable amount of time are that essentially all of the constraints are linear (though there are exceptions), and that the integer variables appear only in the objective function and the budgetary constraints—not in the dam revenue equation nor in the recreational, municipal, industrial, and physical constraints placed on the river basin.

As discussed in Section 2.3.6, most of the significant features of a river basin are included in the model, although some of the more nebulous factors and those for which realistic parameters are difficult to obtain have been omitted. The major simplification has been the elimination of a bay or estuary at the termination of the river and the elimination of water quality factors. However, water quality is taken into consideration in Chapter 6. In the next chapter we examine the strategy for optimizing the model and provide the details of an algorithm that can be used to solve the problem posed in this chapter in a reasonable amount of computer time.

Chapter 3

A PROCEDURE FOR SOLVING THE OPTIMAL EXPANSION PROBLEM

In Chapter 2 the problem of the optimal expansion of a water resources system has been formulated in a way designed to facilitate solution of the problem. More complicated problem statements could be written, of course, but the existing optimization methods that can be applied to solve such problem statements are inadequate either because they do not work or because they take an inordinate amount of computer time or storage. In this chapter we shall describe a strategy for solving the problem posed in Chapter 2.

The problem as summarized in Section 2.4 will be designated for ease of description as Problem I. As it stands it comprises the objective function [expression (2.1)] and various types of constraints [expressions (2.2)–(2.17)], and is a 0–1 mixed integer programming problem. Consequently, all the feasible solutions will contain a mixture of integer and noninteger variables; the integer variables are restricted to the values 0 or 1. We want to specify (1) if and when each dam should be built, and also (2) a sequence of reservoir releases such that the objective function is maximized.

Several nonlinear terms appear in the problem statement that prevent the use of linear integer programming as a tool for solution. Note that constraints (2.7) and (2.10) are nonlinear because of the interaction between $S_{i+1,jt}$ and Q_{imt}. The objective function is nonlinear in two respects: (1) the terms containing the double sum of X_{ijt} are nonlinear, and (2) interaction takes place between the pairs of variables β_{jt} and X_{ijt}, λ_{jt} and C_{jt}, and σ_{jt} and K_{jt}. Because of the nonlinear terms in the problem statement and because of the discreteness of several of the independent and dependent variables only a limited number of solution techniques can be considered for the solution of Problem I.

Possible techniques of solution include the generalized Lagrange multiplier technique [Everett, 1963; Kaplan, 1966], dynamic programming [Bellman and Dreyfus, 1962], and mixed variable programming [Benders, 1962], but each method is rendered ineffective because of some factor or characteristic of Problem I.

The difficulties with the application of the generalized Lagrange multiplier technique are twofold:

1. First, the method requires that the alternative new projects be independent. In Problem I the alternative projects are independent with respect to cost factors and required investment but are interrelated with respect to benefits.
2. Second, the method may fail to provide an optimal solution [Cord, 1964; Weingartner, 1966].

Dynamic programming might be deemed to be suitable to solve Problem I, but previous experience has shown that the computer core storage requirements would be prohibitively large because of the large number of state variables involved in Problem I. Note that:

1. When the problem has more than one constraint, the number of state variables increases.
2. Because of the physical configuration of the problem, a dynamic programming formulation would include converging and diverging branches, with a consequent increase in the number of state variables.

For problems involving a large number of constraints Dantzig [1957] and Nemhauser and Ullmann [1969] reported that dynamic programming was not an efficient tool. Swanson [1970] reached the same conclusion for water resource systems involving several state variables and more than two stages.

A third possible method of optimization is Benders' [1962] algorithm for mixed variable programming problems. This algorithm requires that the objective function be separable with respect to the integer and continuous variables, i.e., the two sets of variables must be capable of being linearly summed. However, there is interaction among variables in the objective function so that Problem I does not meet the condition of separability.

In the Texas Water Development Board Study [1969] it was decided to develop a screening technique that would be able to reduce drastically the numbers of alternatives to be considered. This technique, which utilized a variety of optimization routines to find "near optimum" solutions, embodied in four major phases: (1) initial element sizing and reservoir operating rules using an optimal allocation program, (2) initial screening of development plans using a simulation program, (3) secondary screening of development plans using a detailed simulation program, and (4) final screening of development plans using the allocation program. Although the above approach did not guarantee a minimum-cost solution, it did permit the planner to inject his judgment and experience into the screening process to approach the minimum-cost solution as closely as he desired. It also permitted him to eliminate illogical results and provided him with an opportunity to integrate into the decision-making process those considerations that could not otherwise be expressed in quantitative terms. However, since we are interested in the *optimal* expansion of a water resources system and seek an extremum, some other technique of solution must be applied.

3.1. Strategy of the Optimization Algorithm

An algorithm that takes advantage of the discreteness of the variables involved in Problem I and views it as a combinatorial problem can be

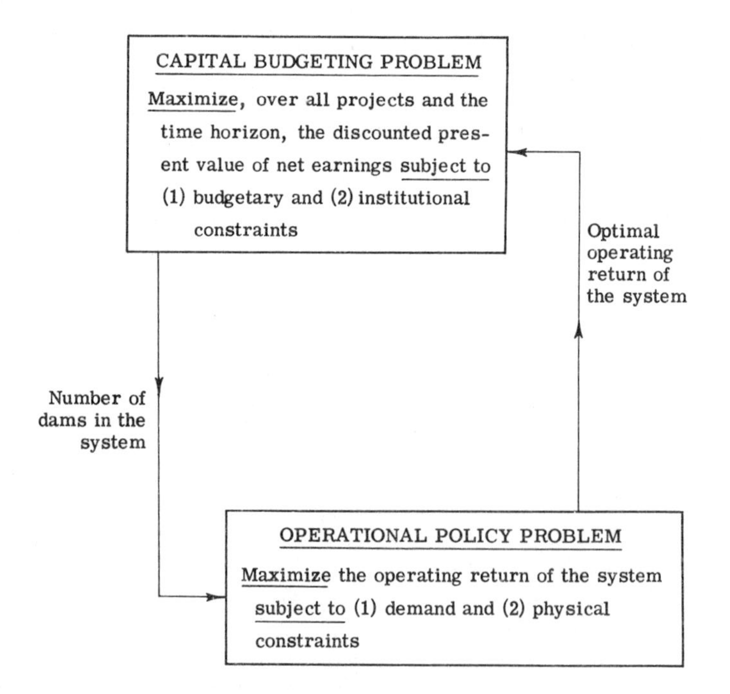

Fig. 3.1 Decomposition of the problem of the optimal expansion of a water resources system.

successfully employed to find a solution. The essence of the strategy of the algorithm is to (1) decompose Problem I into the set of all feasible combinations (termed Problem II), and (2) consider the economic return for each combination (termed Problem III). The combination of (1) and (2) with the best return is necessarily the optimum solution for Problem I.

Problem II is a capital budgeting (CB) problem and is concerned with the *allocation of capital* among the new dams to be built. Problem II can be stated as follows:

Maximize the objective function of Problem I *subject to* constraints (2.2)–(2.6) of Problem I.

Problem III is an operational policy (OP) problem and is concerned with the *allocation and flow of water*. Problem III can be stated as follows:

Maximize the revenue X_{ijt} in Eq. (2.7) of Problem I for all j, $j = 1, 2, \ldots, M$, *subject to* constraints (2.8)–(2.17) of Problem I.

The two problems are interconnected because X_{ijt} in the CB problem is determined only by obtaining the *optimal solution* of the OP problem, and the total number of dams in the OP problem is the *optimal solution* of the CB problem (see Fig. 3.1). We shall examine methods of solving Problems II and III separately.

3.2. The Capital Budgeting Problem (Problem II)

The CB problem is solved by using a branch and bound algorithm. Before describing the algorithm itself, a few remarks are pertinent concerning the branch and bound method as an optimization technique.

3.2.1. The Branch and Bound Method As an Optimization Tool

In optimization a branch and bound algorithm (BBA) comprises a heuristically structured search of the space of all feasible solutions. A number of BBA's have been proposed to solve a wide variety of combinatorial problems [Golamb and Baumert, 1965; Little *et al.*, 1963]. Hillier and Lieberman [1967, pp. 565–570] provide a brief, simple summary of the technique. By combinatorial problem we mean an optimization problem that has some objective function $f(\mathbf{x})$ to be minimized or maximized subject to a set of constraints, with the extremum to be established by the assignment of values to the set of variables $\mathbf{x}$. Considerable flexibility exists in both the nature of the objective function and the constraints. Combinatorial problems with nonlinear, discontinuous, discrete, and even nonmathematically defined objective functions can be solved as long as $f(\mathbf{x})$

is uniquely determined by $\mathbf{x}$. A particularly important feature of the branch and bound method is that an optimal solution to a problem can be obtained with less than complete enumeration of all the possible solutions.

Three typical examples of combinatorial problems that have been solved by BBA's (as well as by other methods, of course) are the following.

The Traveling Salesman Problem

The problem is to assign values of 0 or 1 to variables x_{ij}, where x_{ij} is 1 if the salesman travels from city i to city j and 0 otherwise. The constraints in the problem are that the salesman must start at a particular city, visit each of the other cities only once, and return to the original city. Some cost (here distance) c_{ij} is associated with traveling from city i to city j, and the objective function is to

Minimize the total cost of the trips to each city

$$f(\mathbf{x}) = \sum_{i=1}^{n} \sum_{j=1}^{n} c_{ij} x_{ij} = \sum_{i=1}^{n} \mathbf{c}_i^{\mathrm{T}} \mathbf{x}_i$$

subject to

$$\sum_{i=1}^{n} x_{ij} = 1 \qquad \sum_{j=1}^{n} x_{ij} = 1$$

where

$$\mathbf{c}_i^{\mathrm{T}} = [c_{ij}, \ldots, c_{in}] \qquad \mathbf{x}_i^{\mathrm{T}} = [x_{ij}, \ldots, x_{in}]$$

The Machine Job-Shop Scheduling Problem

The problem is to assign integer values to variables x_{ij}, where x_{ij} is the starting time of job i on machine j; $j = 1, \ldots, n$. The constraints in the problem are that a job cannot be processed on machine n before it has been completed on machine $(n - 1)$, and it cannot be processed on machine $(n - 1)$ before it has been completed on machine $(n - 2)$, and so forth. Given the time t_{ij} that it takes to complete the work of job i on machine j, the problem is to schedule the jobs on each machine so that the total time for the completion of all the jobs is a minimum. The objective function is

$$f(\mathbf{x}) = \max_{i}(x_{in} + t_{in})$$

Because $x_{in} + t_{in}$ is the time at which job i is completed on machine n, the maximum of these numbers is the time at which the latest job is completed. It is that time which is to be minimized.

Because the set of all feasible combinations termed Problem II in Section 3.1 includes aspects of both The Traveling Salesman Problem and The Knapsack Problem, the latter problem is also of interest.

The Knapsack Problem

The Knapsack Problem has two different aspects: (1) if a given space is to be packed with items of different value and volume, the objective is to choose the most valuable packing; or (2) if a given item is to be divided into portions of different value, the objective is to find the most valuable division of the item. A formal statement of the problem is

Maximize

$$f(\mathbf{x}) = \sum_{j}^{n} a_j x_j$$

subject to

$$\sum_{j}^{n} b_j x_j \leq L \qquad x_j = \{0, 1\}$$

The a_j are positive numbers; the b_j and L are positive integers.

Other examples of branch and bound problems are plant location [Efroymson and Ray, 1965; Davis and Ray, 1969] and mixed integer linear programming [Land and Doig, 1960].

In essence the procedure in a BBA is to repeatedly partition the space of all the feasible solutions into smaller and smaller subsets. An upper bound in the case of maximization (or lower bound for minimization) is computed for the value of the objective function for each subset. The partitioning, also termed branching, is carried out so that the subsets are mutually exclusive and each feasible solution belongs to only one subset. After the initial branching, branching is continued, but those subsets with an upper bound that is less than the bound of a known feasible solution are not partitioned any further and are excluded from further consideration. Otherwise, branching continues repeatedly until a feasible solution is obtained that has a value for the objective function greater than the upper bounds of all the other remaining subsets. The branch and bound method solves a difficult problem by using well known techniques to solve a series of easier problems.

To see how this decomposition takes place, suppose that we want to solve the following problem.

Problem 0

Maximize

$$f^{(0)}(\mathbf{x})$$

subject to

$$g_i^{(0)}(\mathbf{x}) \geq 0 \qquad i = 1, \ldots, m; \quad \mathbf{x} \in E^n, \quad \mathbf{x} \geq \mathbf{0}$$

Let us replace Problem 0 with an easier problem, Problem 1, that "bounds" Problem 0 in the following sense [Lawler and Wood, 1966]:

There exists at least one optimal feasible solution $\mathbf{x}^{(0)}$ of Problem 0 such that $\mathbf{x}^{(0)}$ is feasible for Problem 1 and $f^{(1)}(\mathbf{x}^{(0)}) \geq f^{(0)}(\mathbf{x}^{(0)})$. (B)

Here $f^{(1)}(\mathbf{x}^{(0)})$ denotes the objective function for Problem 1, a problem of similar form to Problem 0 but not necessarily having the same list of constraints. Furthermore, if we find an optimal feasible solution $\mathbf{x}^{(1)}$ to Problem 1, it can be shown that

If $\mathbf{x}^{(1)}$ satisfies the optimality conditions that (a) $\mathbf{x}^{(1)}$ is a feasible solution to Problem 0, and (b) $f^{(1)}(\mathbf{x}^{(1)}) = f^{(0)}(\mathbf{x}^{(1)})$, then $\mathbf{x}^{(1)}$ is an optimal solution to Problem 0 as well.

Because it may not be easy for $\mathbf{x}^{(1)}$ to satisfy requirements (a) and (b), it generally proves better to replace Problem 1 by a set of problems $P = \{2, 3, \ldots\}$ that bound Problem 0 in the sense that they jointly satisfy the following bounding property [Lawler and Wood, 1966]:

There exists at least one optimal solution $\mathbf{x}^{(0)}$ of Problem 0 such that $\mathbf{x}^{(0)}$ is feasible for at least one problem j of the set P, and $f^{(j)}(\mathbf{x}^{(0)}) \geq f^{(0)}(\mathbf{x}^{(0)})$. (B′)

If we find an optimal solution $\mathbf{x}^{(j)}$ for each of the j problems in the set P, and define $\mathbf{x}^{(k)}$ such that

$$f^{(k)}(\mathbf{x}^{(k)}) = \max_{j \in P}[f^{(j)}(\mathbf{x}^{(j)})]$$

it can be shown that $\mathbf{x}^{(k)}$ is an optimal solution to Problem 0:

If $\mathbf{x}^{(k)}$ satisfies the optimality conditions that (a′) $\mathbf{x}^{(k)}$ is a feasible solution to Problem 0, and (b′) $f^{(k)}(\mathbf{x}^{(k)}) = f^{(0)}(\mathbf{x}^{(k)})$, then $\mathbf{x}^{(k)}$ is an optimal solution to Problem 0.

To continue in this vein, if $\mathbf{x}^{(k)}$ does not satisfy the conditions (a′) and (b′), again we replace one of the problems in the bounding set P by a new set of problems. Suppose we replace Problem k by a set of bounding problems $P_k = \{P_{k1}, P_{k2}, \ldots\}$. In addition to requiring that the union of the set P, less Problem k, with the set P_k satisfy the bounding condition B′, two convergence conditions are imposed.

Weak Convergence

For each problem j in the set P_k, either $\mathbf{x}^{(k)}$ is infeasible for j or else $f^{(j)}(\mathbf{x}^{(k)}) < f^{(k)}(\mathbf{x}^{(k)})$. (C)

Stronger Convergence

For each problem j in the set P_k and each feasible solution $\mathbf{x}$ to problem k, either $\mathbf{x}$ is nonfeasible for problem j or $f^{(j)}(\mathbf{x}) < f^{(k)}(\mathbf{x})$. (C′)

Of course, conditions C and C′ are not sufficient to guarantee that repeated partioning of the set P will yield an optimal solution to Problem 0 with a finite amount of computation, but for reasonably sized problems and large digital computers the lack of a guarantee does not prove to be much of a handicap.

Figure 3.2 assists in the interpretation of the branch and bound procedure outlined above. Figure 3.2 represents a tree. Each node of the tree corresponds to a problem j.

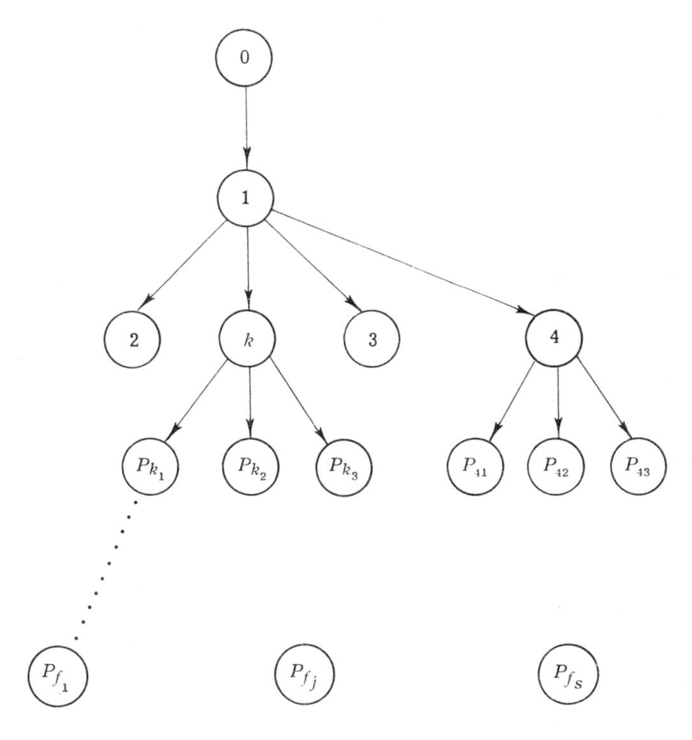

Fig. 3.2 Tree representation of the branch and bound procedure. The final nodes represent single solutions.

Problem j

Maximize

$$f^{(j)}(\mathbf{x})$$

subject to

$$g_i^{(j)}(\mathbf{x}) \geq 0 \qquad i = 1, 2, \ldots$$

This problem is similar to Problem 0 except that the specific constraints and the total number of constraints included in each of the problems may differ. The branches in the tree lead from Problem 0 to the subproblems in the bounding sets. At any stage of the procedure the "leaves" on the tree represent the current union of the sets of bounding problems.

If we let $\hat{\mathbf{x}}$ denote the best feasible solution that has been found at any stage of the calculations, that is, $f^{(0)}(\hat{\mathbf{x}})$ is the largest value (for maximization) for any bounding problem j (with $\mathbf{x}^j \neq \hat{\mathbf{x}}$), if $f^{(0)}(\hat{\mathbf{x}}) \geq f^{(j)}(\mathbf{x}^{(j)})$, it is clear that problem j can be removed from the set P without affecting the bounding condition B′. We say that there is associated with each node j of the tree a bound $f^{(j)}(\mathbf{x}^{(j)})$, and that any leaf node of the tree whose bound is greater than $f^{(0)}(\hat{\mathbf{x}})$ is *active*; if the bound is equal to or less than $f^{(0)}(\hat{\mathbf{x}})$, the leaf can be *terminated*, that is, the bound need not be considered in any further computations. The general procedure in branching and bounding is to develop the tree until every leaf can be terminated, and to work out a suitable strategy so that not too many leaves will have to be decomposed into subproblems. Appropriate strategies include rules for deciding which of the active bounding problems is to be selected for further branching, together with methods for deciding how to formulate the new bounding problems.

3.2.2. Little's Branch and Bound Algorithm

The previous section dealt with the general characteristics of the BBA. This section describes Little's branch and bound algorithm (LBBA), which was developed especially to solve the traveling salesman problem. A legitimate initial lower bound (L_1) for LBBA is the sum of the minimum elements of the rows of the $\mathbf{C}$ matrix (whose elements are c_{ij}) plus the sum of the minimum elements of the columns of the matrix *after it has been modified* by subtracting from each element the value of the lowest element of its own row. This process is called the row and column reduction of the matrix. For the other nodes, the distance matrix $\mathbf{C}$ is modified to include (or exclude) the arcs associated with the node, and the bounds are equal to L_1 plus the value obtained by a row and column reduction.

Two subproblems are created at each branching step corresponding to $X_{ij} = 0$ and $X_{ij} = 1$. At each branching step the arc (i, j) is chosen in such a way that the problem corresponding to $X_{ij} = 0$ will yield as large a bound as possible. Since the algorithm will always first branch to a node having a lower bound, this heuristic rule plus the rules mentioned in the following paragraph will tend to make the first feasible solution the optimal one.

The branching technique is as follows:

1. At any stage of computation, the distance matrix will contain one or more zeros. All the arcs corresponding to the zero values are candidate arcs in the optimal solution.

2. Define a variable (θ_{ij}) that measures the change in the value of the bound for *not including* arc (i, j) (whose $c_{ij} = 0$) in the final solution:

$$\theta_{ij} = \alpha_i + \beta_j$$

where α_i is the second lowest element of row i, and β_j the second lowest element of column j. Calculate θ_{ij}.

3. For the next branching stage pick the arc (i, j) that corresponds to the *maximum* value of θ_{ij}.

All the concepts of this section are illustrated in the following section, where the LBBA is used to solve a simple traveling salesman problem.

3.2.3 Example of Application of Little's Branch and Bound Algorithm

(a) PROBLEM STATEMENT

A traveling salesman must visit four cities designated 1, 2, 3, and 4 (see Fig. 3.3). The 4×4 symmetric matrix $\mathbf{C} = [\mathbf{c}_1 \ldots \mathbf{c}_n]^T$ whose elements are c_{ij} with the leading diagonal elements equal to infinity gives the distances (in miles) to be traveled between each pair of cites. What route should the salesman choose to minimize the total distance traveled in visiting the four cities and returning to his original starting point?

To prohibit travel in any arc (i, j), the symbol $c_{ij} = \infty$ is used. In matrix 1 all the leading diagonal elements have been made equal to infinity to avoid trips in the (i, i) arcs.

(b) FIRST ROW AND COLUMN REDUCTION

A row and column reduction of matrix 1 gives a lower bound for node P_1 and yields matrix 2, as shown in Fig. 3.4. The lower bound for node P_1 is $L_1 = 286$.

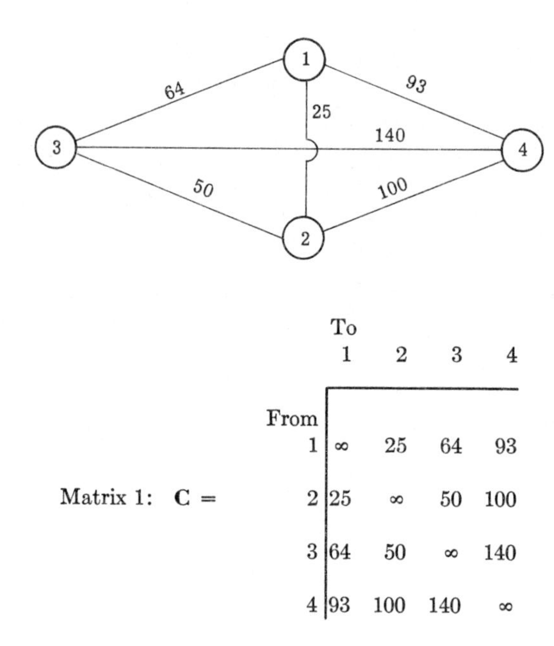

Fig. 3.3 The relative locations of cities 1, 2, 3, 4. Intercity distances are shown in miles; e.g., $c_{34} = 140$ mi.

(c) FIRST BRANCHING SEQUENCE

In matrix 2 there are six elements (whose values of c_{ij} are equal to zero) that could be considered as the first arc of the branching sequence. Table 3.1 gives the values of θ_{ij} for the arcs corresponding to each of these ele-

Table 3.1

Calculation of θ_{ij} for the Zero Elements of Matrix 2

Element (i, j) of matrix 2	α_i	β_j	θ_{ij}
(1, 2)	0	0	0
(1, 4)	0	7	7
(2, 1)	0	0	0
(2, 3)	0	14	14
(3, 2)	14	0	14
(4, 1)	7	0	7

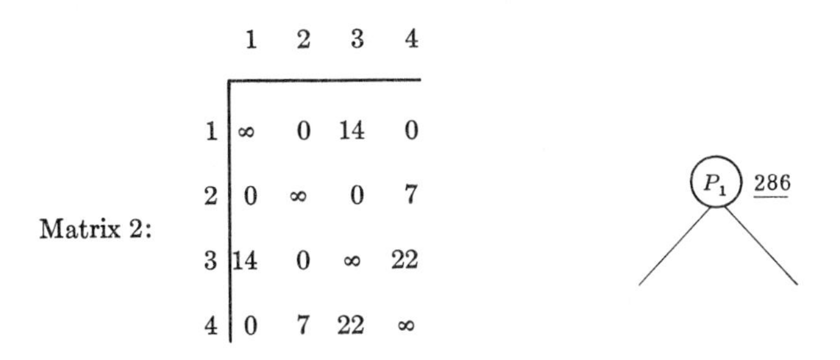

Matrix 2:

	1	2	3	4
1	∞	0	14	0
2	0	∞	0	7
3	14	0	∞	22
4	0	7	22	∞

Fig. 3.4 Starting the solution tree.

ments. Arc (2, 3) or arc (3, 2) has the highest value of θ_{ij}, and according to rule (3) of Section 3.2.2 either arc could be included in the first branching sequence. Arbitrarily, arc (3, 2) was chosen for the first branching sequence. The bound for excluding arc (3, 2) [denoted by $(\overline{3, 2})$] is equal to $286 + 14 = 300$, a value computed by making the element (3, 2) equal to infinity and continuing with row and column reduction. To calculate the bound for including arc (3, 2) it is necessary to change matrix 2 as follows:

1. Element (3, 2) is made equal to infinity to avoid considering it again, and element (2, 3) is made equal to infinity to avoid the subtour (2–3–2).
2. Elements (3, 1), (3, 4), (1, 2), (4, 2) are made equal to infinity since we can only *arrive* at city 2 from city 3 or *depart* for city 2 from city 3. In effect, row 3 and column 2 are eliminated from matrix 2, and matrix 3 results.

Table 3.2

Calculation of θ_{ij} for the Zero Elements of Matrix 4

Element (i, j) of matrix 4	α_i	β_j	θ_{ij}
(1, 3)	0	8	8
(1, 4)	0	7	7
(2, 1)	7	0	7
(4, 1)	8	0	8

Matrix 3:

	1	3	4
1	∞	14	0
2	0	∞	7
4	0	22	∞

A row and column reduction reduces the value of the elements of column 3 in matrix 3 by 14, thereby giving a bound for arc $(3, 2) = 286 + 14 = 300$ and the new matrix 4; see Fig. 3.5.

(d) SECOND BRANCHING SEQUENCE

In matrix 4 there are four elements that could be considered for the second arc of the branching sequence. Table 3.2 gives the value of θ_{ij} for each of the elements.

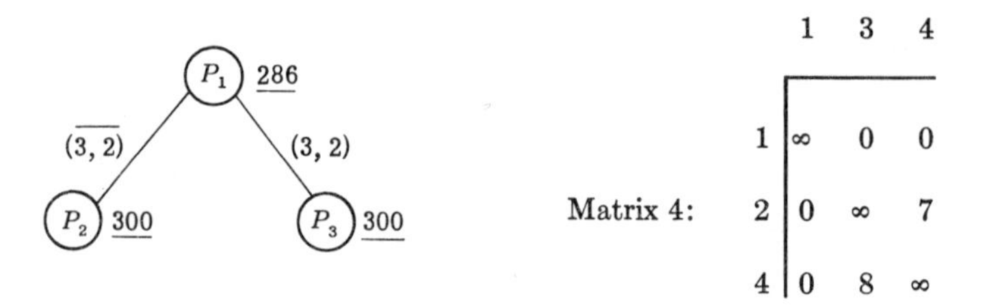

Fig. 3.5 The solution tree after branching sequence 1. Underlined numbers are bounds. Ordered pairs indicate whether or not arc is included in a particular branch.

Arc (1, 3) is one of the set of highest values of θ_{ij} and will be chosen as the second arc in the branching sequence. The bound for $(\overline{1, 3})$ is $300 + 8 = 308$. The bound for (1, 3) is calculated (1) by removing row 1 and column 3 from matrix 4 to give matrix 5, (2) making element (2, 1) equal to ∞ to avoid the sub-tour 1-3-2-1 and (3) carrying out a row and column reduction of matrix 5.

Matrix 5:

	1	4
2	∞	7
4	0	∞

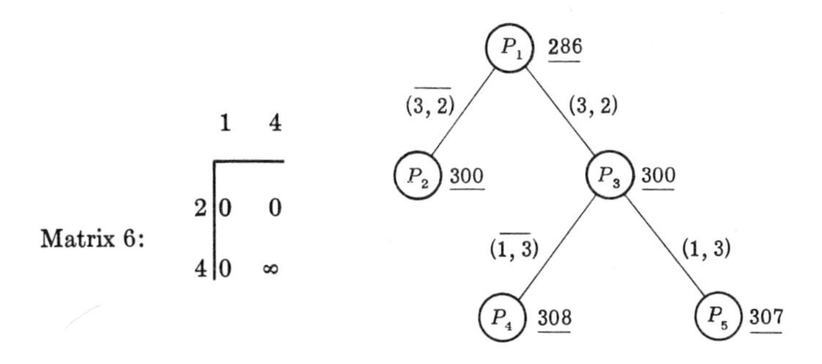

Fig. 3.6 The solution tree after branching sequence 2.

The row and column reduction of matrix 5 reduces the value of the elements of column 4 by 7, thereby giving a bound for arc (1, 3) = 300 + 7 = 307 and the new matrix 6; see Fig. 3.6.

(e) THIRD BRANCHING SEQUENCE

In matrix 6 there are two elements that could be considered for the third branching sequence. Table 3.3 gives the value of θ_{ij} for each of the elements.

The bound for arc (2, 4) is calculated by removing row 2 and column 4 from matrix 6 to give the element (4, 1), which is equal to zero. Thus the bound for arc (2, 4) = 307 + 0 = 307. Also the bound for arc (4, 1) = 307 + 0 = 307. Thus we have a feasible solution to the traveling salesman problem, namely the sequence (3, 2), (1, 3), (2, 4), (4, 1), or rearranging, (1, 3), (3, 2), (2, 4), (4, 1). The length of the route is 307 mi; Fig. 3.7 shows the solution tree.

Table 3.3

Calculation of θ_{ij} for the Zero Elements of Matrix 6

Element $(i\ j)$ of matrix 6	α_i	β_j	θ_{ij}
(2, 4)	∞	∞	∞
(4, 1)	∞	∞	∞

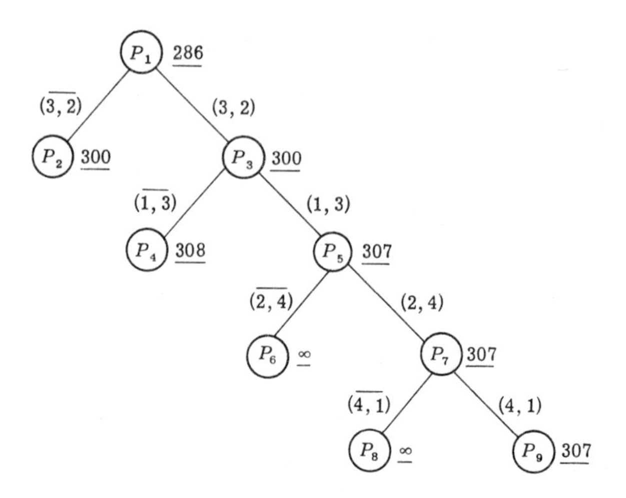

Fig. 3.7 The solution tree after branching sequence 3.

(f) BACKTRACKING SEQUENCE

We have found a feasible solution to the traveling salesman problem, and we now search for those nodes that have bounds lower than the feasible solution. Only node P_2 [corresponding to omitting arc (3, 2)] has a bound lower than the feasible solution. The backtracking proceeds as follows:

1. In matrix 1 make element $(3, 2) = \infty$.
2. Carry out the usual row and column reduction on matrix 1.
3. Proceed in exactly the same manner as described in branching sequences 1 through 3.

One solution is found that has the same value as the first feasible solution, 307, namely the route (1, 4), (4, 2), (2, 3), (3, 1), and is the same route as shown by the tree in Fig. 3.1 except in the opposite direction. All the remaining nodes (P_4, P_6, P_8) had bounds *at least as large* as the value of the first feasible solution. Thus the first feasible solution is an optimal solution of the traveling salesman problem.

3.2.4. Branching and Bounding for the Capital Budgeting Problem

We turn now to consideration of the specific rules for branching and bounding that can be established to solve the capital budgeting problem (Problem II). These rules represent an extension of the work of Little *et al.* [1963] and Lesso [1969]. Figure 3.8 represents a decision tree that shows the set of all feasible combinations of new dams in Problem II. The vertical axis corresponds to the time variable, and the horizontal axis corresponds to the choices that can be made in any given year. In drawing the tree we have observed the previously mentioned assumption that at most only one new facility may be introduced in any year.

Each node signifies a new policy decision: that is, node $N + 1$ of year t represents the proposed building of dam number $N + 1$ in year t. Each arc signifies the continuation of policy from year to year: that is, an arc from node $N + 1$ in year t to node $N + 2$ in year $t + 1$ signifies that $N + 1$ dams have been built at the end of year t and that the N + 2nd dam will be built in the year $t + 1$.

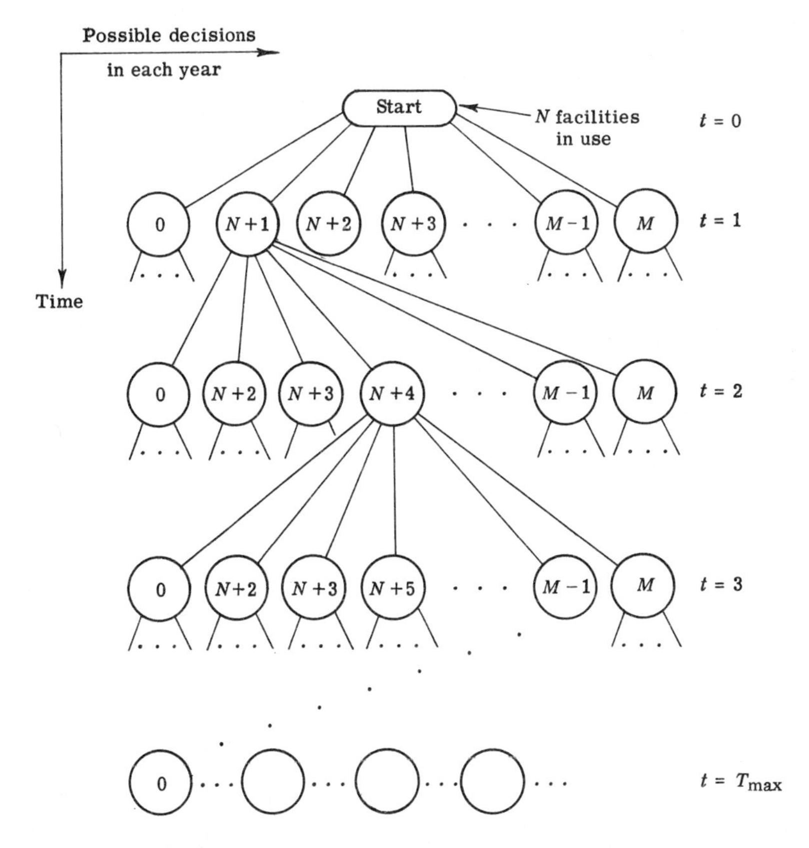

Fig. 3.8 Decision tree representation of Problem II. Each node refers to a possible decision.

(a) BRANCHING PROCEDURE

At time $t = 0$ there are N facilities in use. At time $t = 1$ it may be decided to introduce no new facility (node 0), facility $N + 1$, or any one of the other possible facilities $(N + 2, N + 3, \ldots, M)$. At time $t = 2$, for each one of the alternatives in period 1 (except 0), there are $M - N$ alternatives corresponding to each of the nodes of $(t = 1)$. The alternatives are (1) to introduce no new facility (node 0), or (2) to introduce a new facility $N + 1, N + 2, \ldots, M$, except that no new facility may be repeated. The structure continues until all possibilities are exhausted or until period $T_{\max}$ is reached.

(b) CALCULATION OF THE BOUNDS

A bound B_{jt} is associated with each node j of year t. It is the upper limit on the economic return from the use of the dam(s) associated with the node added to (1) the cumulative net return CR_{t-1} from the system to the end of year $t - 1$ plus (2) the return TR_t from operating the system as it exists at the beginning of year t for the years t through $T_{\max}$. CR_{t-1} and TR_t are defined as follows:

$$CR_{t-1} = \sum_{t=1}^{T-1} [\hat{X}_t/(1 + r)^t] - \sum_{t=1}^{T-1} [1/(1 + r)^t] \sum_{j=N+1}^{M} \lambda_{jt} C_{jt} \qquad (3.1)$$

$$TR_t = \hat{X}_t \sum_{t=t}^{T_{\max}} [1/(1 + r)^t] \qquad (3.2)$$

The bound B_{jt} is calculated by assuming that:

1. The dam j corresponding to the given node is introduced and operated at maximum efficiency for every time period, from the time of its introduction into the system through all the remaining time periods. This assumption leads to the net operational return for reservoir j if introduced at the beginning of year t and operated at maximum efficiency for the years t to $T_{\max}$,

$$OR_{jt} = (MR_j) \sum_{t=t}^{T_{\max}} [1/(1 + r)^t] - \lambda_{jt} C_{jt} [1/(1 + r)^t] \qquad (3.3)$$

where MR_j is the maximum annual return from operating reservoir j.

2. In the following time period the best remaining dam (i.e., the dam with the highest bound) is introduced into the system and is operated at maximum efficiency until the end of the time horizon $T_{\max}$, to give $OR_{j',t+1}$, and so on for all the remaining dams and time periods.

Thus B_{jt} is defined by

$$B_{jt} = CR_{t-1} + TR_t + OR_{jt} + OR_{j',t+1} + OR_{j'',t+2} + OR_{j^m,T_{max}} \quad (3.4)$$

where $j', j'', \ldots, j^m$ are the ranking of the OR_{jt} for the remaining alternatives in the respective time periods.

We should note that when B_{jt} is calculated, all the projects incompatible with project j are not considered. For example, if project $N + 1$ is incompatible with two other projects $N + 2$ and $N + 3$ [refer to constraint (2.4) of the problem formulation in Chapter 2], B_{jt} is calculated without consideration of project $N + 1$ or $N + 2$.

(c) PROCEDURE FOR SEARCHING THROUGH THE TREE

For month (period) 1 the bounds for every node are calculated and ranked in the order of their maximum return. The node with the highest bound that meets the budgetary contraint is selected and the operational policy (OP) problem is solved for 12 periods (1 year) with a new dam added at the beginning of year 1. The OP problem is also solved for 12 periods (1 yr) for the case in which *no* dam is added. *If and only if* the solution of the former OP problem gives a higher return than the solution of the latter OP problem should a new dam be constructed in year 1.

Branching takes place from the node corresponding to the higher OP return. All other nodes with bounds *lower* than the return from the branching node are eliminated.

This procedure is repeated for each subsequent year. On termination of branching at T_{max} or when all the dams have been built that can be, the next step is to backtrack up the tree to examine any remaining intermediate nodes and to branch from them until they are either eliminated or replace the current feasible solution. The optimal solution is found when all the feasible nodes have either been searched or eliminated.

3.3. The Operational Policy Problem (Problem III)

Once it has been decided that a new dam may be added, its feasibility must be tested with respect to the river basin model. The water resources system can be regarded as a three-dimensional network one with one axis corresponding to the variable time and the other two axes corresponding to the physical flow in two-dimensional space through the system for each month. Figure 3.9 is a sketch of the information flow for a typical year of 12 periods (months). Any of the optimization techniques described in

Section 1.3 can be employed in solving the OP problem. If the objective function and/or constraints are nonlinear, a nonlinear programming algorithm must be used. However, in view of the amount of computer time that is currently required to solve a single nonlinear programming problem, the

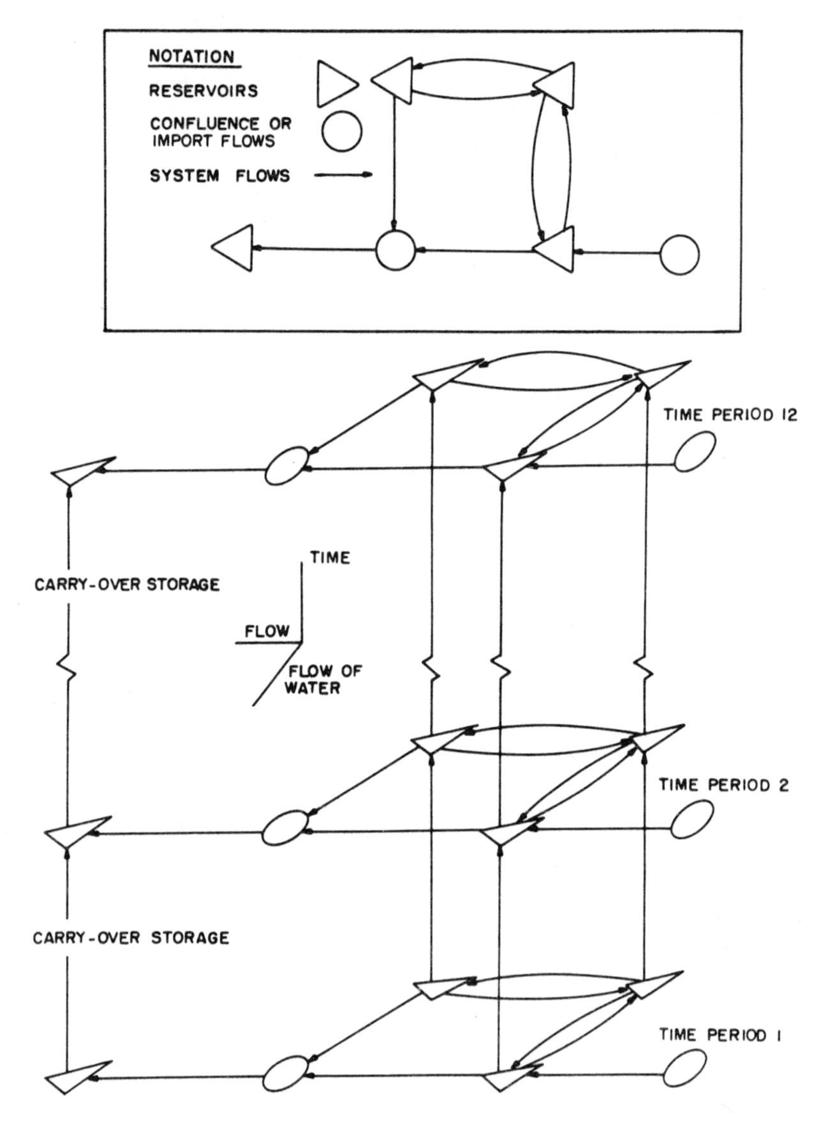

Fig. 3.9 Network representation of a 12-period (one year) OP problem.

Table 3.4

Notation Used in Section 3.3

Symbol	Significance	Units
A	A subset of ordered pairs from a collection of N elements	
b_{ij}	Benefit of passing one unit of flow through the arc (i, j)	\$/acre-ft
c_{ij}	Cost of passing one unit of flow through the arc (i, j)	\$/acre-ft
f_{ij}	Flow in the arc (i, j)	acre-ft
G	A directed network $\{N; A\}$	
l_{ij}	Lower flow capacity of arc (i, j)	acre-ft
N	A collection of elements	
q_{ij}	Total cost to the system-consumer and distributor of transporting one unit of flow from node i to node j	\$/acre-ft
u_{ij}	Upper flow capacity of arc (i, j)	acre-ft
δ_{ij}	Marginal value of decreasing l_{ij} by one unit	\$/acre-ft
γ_{ij}	Marginal value of increasing u_{ij} by one unit	\$/acre-ft
π_j	Price of one unit of water at the node (j)	\$/acre-ft

feasibility of applying nonlinear programming to the OP problem is poor for realistically sized water resources systems.

The OP problem sketched in Fig. 3.9 is in the form of a network flow problem. Fulkerson [1961] developed the out-of-kilter algorithm to solve this type of problem, provided the objective function is linear or can be approximated by a piecewise linear function (e.g., convex cost functions or concave revenue functions). We will first examine the strategy of the out-of-kilter algorithm and then see how it can be applied to the OP problem.

3.3.1. *The Out-of-Kilter Algorithm*

To clarify the concepts introduced in this section, some new notation is presented in Table 3.4. The out-of-kilter algorithm (OKA) was developed by Fulkerson [1961] to solve the following problem. Define a directed network $G = \{N; A\}$ that consists of a collection of N nodes $\{1, 2, \ldots, N\}$ together with a subset A of the ordered pairs $\{i, j\}$ of elements taken from N. See Fig. 3.10. The pairs $\{i, j\}$ are referred to as arcs. Associated with each arc is an upper (u_{ij}) and a lower (l_{ij}) flow capacity and also the cost

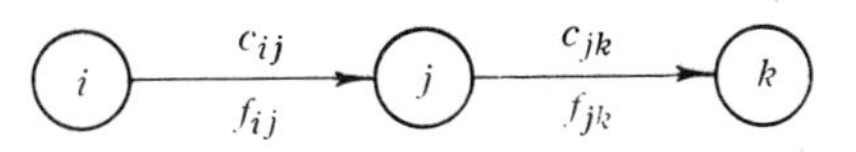

Fig. 3.10 Part of a directed network representing the flow of water.

(c_{ij}) of passing one unit of water through each arc. The flow in arc (i, j) is f_{ij}. The problem is to minimize the cost of passing a feasible flow in the directed network, if one exists.

This is a linear programming (minimum-cost circulation) problem with the special feature that a number of equality constraints exist; each equality constraint corresponds to a mass balance at each node [Eq. (3.7)]. Fulkerson treated the equality constraints in a special way in his algorithm. Computational results for some large-scale problems [Texas Water Development Board, 1969] show the OKA to produce a solution in one twentieth to one fiftieth the time of standard linear programming codes for two reasons: (1) all operations are additive (i.e., no multiplication or division takes place), and (2) no matrix inversion is necessary.

It is convenient to express the problem in linear programming notation so that the relationship between the primal and dual variables may be demonstrated:

Primal Problem

Maximize

$$Z = \sum_i \sum_j b_{ij} f_{ij} \tag{3.5}$$

subject to

$$f_{ij} \leq u_{ij} \qquad \text{for each} \quad (i, j) \in A \tag{3.6}$$

$$\sum_j f_{ij} - \sum_j f_{ji} = 0 \qquad \text{for each} \quad i \in N \tag{3.7}$$

$$-f_{ij} \leq -l_{ij} \qquad \text{for each} \ (i, j) \in A \tag{3.8}$$

$$f_{ij} \geq 0 \qquad \text{for each} \ (i, j) \in A \tag{3.9}$$

Note that maximizing Z is the same as minimizing $\hat{Z} = \sum_i \sum_j c_{ij} f_{ij}$ so that each b_{ij} is a negative coefficient and equals $(-c_{ij})$. From the duality theory of linear programming, there is a dual variable associated with each primal constraint, and a dual constraint associated with each primal variable. Let π_i denote the dual variable associated with the ith primal conservation-of-flow equation [from the set of equations of (3.7)] for each node $i \in N$. Let γ_{ij} denote the dual variable associated with the upper-bound constraint of arc (i, j) from the set of primal inequalities represented by Eq. (3.6). Let δ_{ij} denote the dual variable associated with the lower-bound constraint of arc (i, j) from the set of primal inequalities represented by Eq. (3.8). The mathematical statement of the dual problem is as follows.

Dual Problem

Minimize

$$Y = \sum_A u_{ij}\gamma_{ij} - \sum_A l_{ij}\delta_{ij} \tag{3.10}$$

subject to

$$\pi_i - \pi_j + \gamma_{ij} - \delta_{ij} \geq b_{ij} \quad \text{for each} \quad (i, j) \in A \tag{3.11}$$

$$\gamma_{ij} \geq 0, \quad \delta_{ij} \geq 0 \quad \text{for each } (i, j) \in A \tag{3.12}$$

The π variables are unrestricted in sign, since these dual variables are associated with *equality* constraints in the primal formulation.

At the optimum, the values of the primal and dual objective functions are equal. The relationships between the primal and dual variables that force such an equality have been termed the complementary slackness conditions. These conditions are:

$$\gamma_{ij} > 0 \quad \Rightarrow \quad f_{ij} = u_{ij} \quad \text{for each } (i, j) \in A \tag{3.13}$$

$$\delta_{ij} > 0 \quad \Rightarrow \quad f_{ij} = l_{ij} \quad \text{for each } (i, j) \in A \tag{3.14}$$

$$\pi_i - \pi_j + \gamma_{ij} - \delta_{ij} = b_{ij} \quad \Rightarrow \quad l_{ij} \leq f_{ij} \leq u_{ij} \tag{3.15}$$

where $\Rightarrow$ designates "it follows that."

The OKA is so efficient because it takes advantage of the above relationships and the special structure of the minimum-cost circulation problem in order to examine only a relatively small subset of the primal and dual variables in search of a "best" or optimal value of Eq. (3.5). The efficiency is achieved by defining three quantities

$$q_{ij} = \pi_i - \pi_j - b_{ij} \tag{3.16}$$

$$\gamma_{ij} = \max\,(0, -q_{ij}) \tag{3.17}$$

$$\delta_{ij} = \max\,(0, q_{ij}) \tag{3.18}$$

so that γ_{ij} and δ_{ij} continually satisfy conditions (3.11) and (3.12). Thus the values of π_i may be freely chosen without disrupting dual feasibility. By comparing Eqs. (3.13)–(3.15) with the definitions in Eqs. (3.16)–(3.18), the complementary slackness conditions may be reformulated:

$$q_{ij} < 0 \quad \Rightarrow \quad f_{ij} = u_{ij} \tag{3.19}$$

$$q_{ij} > 0 \quad \Rightarrow \quad f_{ij} = l_{ij} \tag{3.20}$$

$$q_{ij} = 0 \quad \Rightarrow \quad l_{ij} \leq f_{ij} \leq u_{ij} \tag{3.21}$$

These three conditions, along with the conservation of flow to maintain primal feasibility, given in Eq. (3.7), determine the sufficient conditions of optimality for the given problem. The three sufficient conditions may be used to test whether the tentative values of the variables in arc (i, j) are optimal.

If a set of values of π_i and f_{ij} can be found that satisfy Eqs. (3.7) and (3.19)–(3.21) for *each* arc of G, then the values f_{ij} are the solution to the minimum-cost circulation problem. Fulkerson named the OKA by noting that an arc that did not satisfy at least one of the above optimality conditions, Eqs. (3.17)–(3.19), was "out of kilter." The OKA seeks to systematically direct flows and assign values to the π variables so that each arc is "in kilter."

3.3.2. Significance of the Dual Variables

π_i can be considered the price of a unit of the flow commodity at the node i; q_{ij} represents the total cost to the system-consumer and distributor of transporting one unit of flow from node i to node j [Durbin and Kroenke, 1967]; γ_{ij} represents the marginal value of increasing u_{ij} by one unit; and δ_{ij} represents the marginal value of decreasing l_{ij} by one unit.

3.3.3. Possible States of an Arc

Nine mutually exclusive "kilter conditions" are possible for each arc as the algorithm proceeds to seek an optimal solution (Table 3.5). The values

Table 3.5

Possible Kilter Conditions for an Arc

State	q_{ij}	f_{ij}	In kilter?
A	$q > 0$	$f = l$	Yes
B	$q = 0$	$l \leq f \leq u$	Yes
C	$q < 0$	$f = u$	Yes
A_1	$q > 0$	$f < l$	No
B_1	$q = 0$	$f < l$	No
C_1	$q < 0$	$f < u$	No
A_2	$q > 0$	$f > l$	No
B_2	$q = 0$	$f > u$	No
C_2	$q < 0$	$f > u$	No

of q_{ij} and f_{ij} determine whether a given arc is in kilter or, if out of kilter, what changes are needed in the q_{ij} and f_{ij} values to bring it into kilter. The value of q_{ij} is altered by systematically varying the values of π_i and π_j, which can be easily done. However, to change the value of f_{ij} is much more difficult because conservation of flow must be maintained at each node. The OKA changes flow in such a way as to avoid disruption of the conservation of flow.

If flow in the arc (i, j) is to be changed, a path must be found from node j to node i, which, with the inclusion of the (i, j) arc, forms a cycle. Changing the flow in this cycle will maintain conservation of flow at all nodes. The path from node j to node i and the change in flow are chosen in such a way that (1) no in-kilter arc becomes out of kilter, and (2) no out-of-kilter arc becomes more out of kilter. (An arc becomes more out of kilter when its flow is changed in magnitude so as to increase the deviation from the feasible state of the arc. For example, if an arc (i, j) is in state A_1 (i.e., $f_{ij} < l_{ij}$) and its flow is reduced further, then the arc becomes more out of kilter.)

For example, consider the minimum-cost circulation problem represented by Fig. 3.11. The ordered triple $(l_{ij}, u_{ij}, -b_{ij})$ is shown on each arc. The original arc flows and node numbers (π values) appear as underlined numbers. For example, $f_{13} = \underline{2}$ and $\pi_1 = \underline{0}$.

Notice that arc (1, 3) is out of kilter and in state A_2. To bring the arc into kilter the flow must be reduced to the arc lower bound (0) and also conservation of flow in the network must be maintained. This can be achieved by changing flows in the closed path or loop (1, 2), (3, 2), (1, 3). (See Section 3.3.8, iteration 1). Note that the path is not formed by arcs whose flows form a directed graph; the rule to remember for reducing flows in a closed path is to increase flows in forward arcs and to decrease flows in

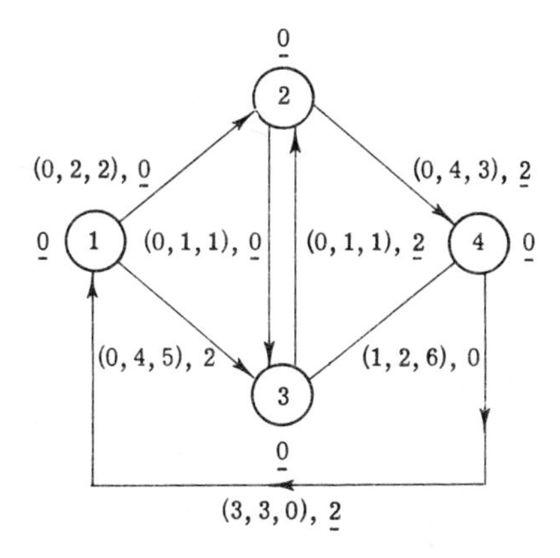

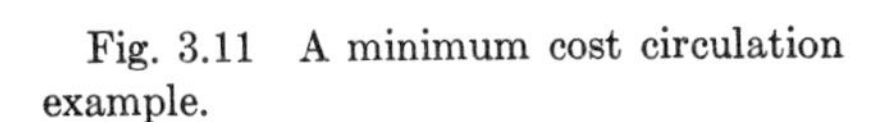
Fig. 3.11 A minimum cost circulation example.

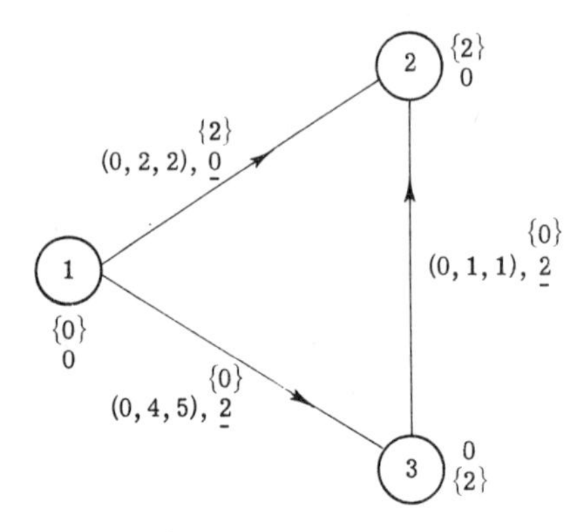

Fig. 3.12 A minimum cost circulation example: flow values and node numbers in iteration 3. For each arc the ordered triple is $(l_{ij}, u_{ij}, -b_{ij})$; π_i and f_{ij} are the node numbers and flow before iteration 1, $\{\pi_i\}$ and $\{f_{ij}\}$ are those in iteration 3.

reverse arcs. Figure 3.12 shows the change in the flows values and node numbers after two complete iterations. Because only a finite number of arcs may be out of kilter, termination of the algorithm is assured. Durbin and Kroenke (1967), Ford and Fulkerson (1962), and Fulkerson (1961) have described a labeling procedure for changing the flow in out-of-kilter arcs. A description of this procedure follows.

3.3.4. Labeling Procedure

1. If an arc (i, j) is found in which the flow is to be *increased* to bring that arc into kilter, then the arc must be in state A_1, B_1, or C_1. Label node j $[i^+, e(j)]$. This signifies that node j may *receive* $e(j)$ units from node i. If the arc is in state A_1, define $e(j)$ to be $(l_{ij} - f_{ij})$; and if the arc is in state B_1 or C_1, define $e(j)$ to be $(u_{ij} - f_{ij})$. These definitions put arc (i, j) into kilter with the fewest changes of the original flow pattern.

2. If the flow in arc (i, j) is to be *decreased*, then the arc must be in state A_2, B_2, or C_2. Label node i $[j^-, e(i)]$, meaning that the flow from node i to node j can be reduced by $e(i)$. If the arc is in state A_2 or B_2, define $e(i)$ to be $(f_{ij} - l_{ij})$; if the arc is in state C_2, define $e(i)$ to be $(f_{ij} - u_{ij})$. An arc that is in any one of the three possible remaining states, A, B, or C is in kilter and its flow should not be changed.

Once an out-of-kilter arc (i, j) has been found, the node i or j is labeled by rule (1) or (2). If the change indicated by the label can be made, the arc can be brought into kilter. However, as discussed previously, a path must be found from j to i [which, with the inclusion of arc (i, j), forms a cycle] so that conservation of flow will not be violated by the proposed flow change. Each node of the path must be labeled so that the needed changes can be accounted for.

3.3.5. Scanned Nodes and the Flow-Augmenting Path

Suppose that arc (i, j) is out of kilter and that node i has been labeled. The aim is to find a flow-augmenting path form i to j in such a way that in-kilter arcs in the path will not be driven out of kilter. Consider each arc that originates or terminates at a labeled node (node i is the first

Table 3.6

Labeling Process Through a Forward Arc

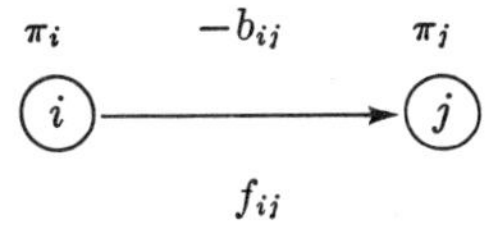

π_i $-b_{ij}$ π_j

$(i) \longrightarrow (j)$

f_{ij}

$$q_{ij} = \pi_i - \pi_j - b_{ij}$$

Suppose that i is labeled $[z^{\pm}, e(i)]$; can j be labeled? (Never label j from i if an increase in the flow f_{ij} will make the arc more out of kilter.)

State of arc (i, j)	q_{ij}	f_{ij}	In kilter?	Can j be labeled?	Why?
A	$q > 0$	$f = l$	Yes	No	Flow increase makes arc out of kilter
B	$q = 0$	$f < u$	Yes	Yes	Flow may be increased to u_{ij}
		$f = u$	Yes	No	Flow cannot be increased
C	$q < 0$	$f = u$	Yes	No	Flow cannot be increased
A_1	$q > 0$	$f < l$	No	Yes	Flow may be increased to l_{ij}
B_1	$q = 0$	$f < l$	No	Yes	Flow may be increased to u_{ij}
C_1	$q < 0$	$f < u$	No	Yes	Flow may be increased to u_{ij}
A_2	$q > 0$	$f > l$	No	No	Flow increase makes arc more out of kilter
B_2	$q = 0$	$f > u$	No	No	Flow increase makes arc more out of kilter
C_2	$q < 0$	$f > u$	No	No	Flow increase makes arc more out of kilter

Summary

Label j $[i^+, e(j)]$.
If $q_{ij} > 0$ and $f_{ij} < l_{ij}$, then $e(j) = \min[e(i), (l_{ij} - f_{ij})]$.
If $q_{ij} \leq 0$ and $f_{ij} < u_{ij}$, then $e(j) = \min[e(i), (u_{ij} - f_{ij})]$.

labeled node), and in each case attempt to label the node at the arc's connecting end. Labeling a node will be possible only if the arc is in one of the allowable states (A_1, B_1, C_1, A_2, B_2, and C_2); the proper label values for flow changes in forward and reverse arcs are given and described in Tables 3.6 and 3.7, respectively. After each incident arc has been considered, the

Table 3.7

Labeling Process Through a Reverse Arc

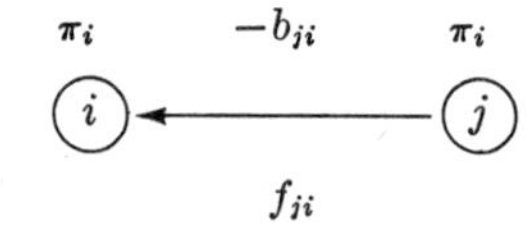

π_i $-b_{ji}$ π_i

(i) ← (j)

f_{ji}

$$q_{ji} = \pi_j - \pi_i - b_{ji}$$

Suppose that i is labeled $[z^{\pm}, e(i)]$; can j be labeled? (Never label j from i if a decrease in the flow f_{ji} will make the arc more out of kilter.)

State of arc (j,i)	q_{ji}	f_{ji}	In kilter?	Can j be labeled?	Why?
A	$q > 0$	$f = l$	Yes	No	Flow decrease would make arc out of kilter
B	$q = 0$	$f \leq u$	Yes	Yes	Flow can be decreased by $f_{ji} - l_{ji}$
		$f = l$		No	Flow cannot be decreased
C	$q < 0$	$f = u$	Yes	No	Flow cannot be decreased
A_1	$q > 0$	$f < l$	No	No	Flow decrease would make arc more out of kilter
B_1	$q = 0$	$f < l$	No	No	Flow decrease would make arc more out of kilter
C_1	$q < 0$	$f < u$	No	No	Flow decrease would make arc more out of kilter
A_2	$q > 0$	$f > l$	No	Yes	Flow may be decreased by $f_{ji} - l_{ji}$
B_2	$q = 0$	$f > u$	No	Yes	Flow may be decreased by $f_{ji} - l_{ji}$
C_2	$q < 0$	$f > u$	No	Yes	Flow may be decreased by $f_{ji} - u_{ji}$

Summary

Label j $[i^-, e(j)]$
If $q_{ji} \geq 0$ and $f_{ji} > l_{ji}$, then $e(j) = \min[e(i), (f_{ji} - l_{ji})]$.
If $q_{ji} < 0$ and $f_{ji} > u_{ji}$, then $e(j) = \min[e(i), (f_{ji} - u_{ji})]$.

given labeled node is marked *scanned.* Now choose a labeled, unscanned node x and repeat the above procedure, attempting to label each node that forms an arc incident to node x. Continue choosing and labeling nodes until j is labeled. If node j is labeled, a *flow-augmenting path* has been found and the flow in the connecting cycle is changed according to the label value. Now arc (i, j) is either in kilter or less out of kilter. Pick an out-of-kilter arc and repeat the procedure.

3.3.6. Nonbreakthrough

However, it is not always possible to pick a flow-augmenting path because the algorithm avoids those arcs for which a change in flow will cause the arc to become (more) out of kilter. In this case, the path would halt at a labeled, scanned node from which no unscanned node could be labeled because of the state of each connecting arc. Such an event is designated as *nonbreakthrough.* The impasse may sometimes be resolved by changing the state of some arc(s). The state of arc (i, j) is uniquely determined by $q_{ij} = \pi_j - \pi_j - b_{ij}$, and the π values (unrestricted dual variables) may be changed without affecting feasibility. At nonbreakthrough there are two sets of mutually exclusive nodes: labeled nodes and unlabeled nodes. The only π values of interest are those that will change the state of arc(s) connecting labeled and unlabeled nodes so that the path may be extended and possibly conpleted. Let X be the set of labeled nodes and $\bar{X}$ be the set of unlabeled nodes. Let M be the set of arc(s) (i, j) originating at a node belonging to X, terminating at a node belonging to $\bar{X}$, and having the property that q_{ij} is positive and f_{ij} is less than or equal to its upper bound. Let $\bar{M}$ be the set of arc(s) (i, j) originating in $\bar{X}$ terminating in X, with q_{ij} negative and f_{ij} greater than or equal to its lower bound.
Thus

$$M = (i,j) \qquad i \in X, \quad j \in \bar{X}, \quad q_{ij} > 0, \quad f_{ij} \leq u_{ij}$$

$$\bar{M} = (i,j) \qquad i \in \bar{X}, \quad j \in X, \quad q_{ij} < 0, \quad f_{ij} > l_{ij}$$

Let

$$D = \min_{M} \{q_{ij}\}, \quad \bar{D} = \min_{\bar{M}} \{-q_{ij}\}, \quad D_f = \min \{D, \bar{D}\}$$

Change the node numbers (π values) of each node in $\bar{X}$ by adding D_f to π_i for each $i \in \bar{X}$, and recompute the state of each arc connecting a node in X to a node in $\bar{X}$. If the state of at least one arc is changed, return to the labeling procedure. If by this process the desired node is labeled, the flow is changed according to the label. All labels are then eliminated, another out-of-kilter arc is chosen, and the operation begins once again. However, if the state of each arc remains unchanged after continued adjustment of the π values, a feasible solution cannot be found and the algorithm terminates. An indication for termination occurs if $D_f = \infty$ at nonbreakthrough.

3.3.7. Summary of the Out-of-Kilter Algorithm

The five principal steps of the OKA may be summarized as follows.

Initial Conditions. Start with a circulation that conserves flow and any set of π values.

Step 1. Find an out-of-kilter arc (i, j). If none, stop. The optimal solution has been found.

Step 2. Determine whether the flow in the arc should be increased or decreased to bring the arc into kilter. If it should be increased, go to step 3; if it should be decreased, go to step 4.

Step 3. Find a path in the network, using the labeling algorithm, from j to i along which the flow can be increased without causing any arc on the path to become (more) out of kilter. If a path is found, increase the flow in the path and also in (i, j). If (i, j) is now in kilter, go to step 1. If it is still out of kilter, repeat step 3. If no path can be found, go to step 5.

Step 4. Find a path from i to j along which the flow can be increased without causing any arc to become (more) out of kilter. If a path is found, increase the flow in the path and decrease the flow in (i, j). If (i, j) is now in kilter, go to step 1. If (i, j) is still out of kilter, repeat step 4. If no path is found, go to step 5.

Step 5. Change the π values and repeat step 2 for arc (i, j), keeping the same labels on all nodes already labeled. If the node numbers become infinite, stop; there is no feasible solution.

3.3.8. Example of Application of the OKA

Figure 3.11 shows a minimum-cost circulation problem that is solved by the OKA. The ordered triple $(l_{ij}, u_{ij}, -b_{ij})$ is shown on each arc. The original arc flows and node number (π values) appear as underlined numbers. For example, $f_{13} = \underline{2}$ and $\pi_1 = \underline{0}$. As initial conditions choose $\pi_i = 0$ for $i = 1, 2, 3, 4$ and $f_{13} = f_{32} = f_{24} = f_{41} = 2$ with $f_{12} = f_{23} = f_{34} = 0$. Note that conservation of flow occurs at every node.

The OKA determines the flow in every arc that gives the minimum-cost circulation. Seven iterations are necessary to obtain the optimal solution; calculations are detailed in the following pages. Figure 3.13 shows

Iteration 1

Arc (i, j)	π_i	$-\pi_j$	$-b_{ij}$	q_{ij}	f_{ij}	State	In kilter?
(1, 2)	0	0	2	2	$0 = l_{12}$	A	Yes
(1, 3)	0	0	5	5	$2 > l_{13}$	A_2	No
(2, 3)	0	0	1	1	$0 = l_{23}$	A	Yes
(2, 4)	0	0	3	3	$2 > l_{24}$	A_2	No
(3, 2)	0	0	1	1	$2 > l_{32}$	A_2	No
(3, 4)	0	0	6	6	$0 < l_{34}$	A_1	No
(4, 1)	0	0	0	0	$2 < l_{41}$	B_1	No

1. Pick the first out-of-kilter arc (1, 3).
2. State of arc is A_2 ; reduce f_{13} to l_{13} .
3. Find path from node 1 to node 3 by labeling procedure.

Labeling procedure	
Node	Label
1	$(3^-, 2)$ (Node 1 is labeled)
2	Cannot be labeled; arc (1, 2) is in kilter
4	Cannot be labeled; flow decrease drives arc (4, 1) more out of kilter

Nonbreakthrough has occurred.

$\dot{X} = \{1\}, \bar{X} = \{2, 3, 4\}, M = \{(1, 2), (1, 3)\}, \bar{M} = \Phi, D = \min\{2, 5\} = 2$, $\bar{D}$ = does not exist, $D_f = 2$.

New π values: $\pi_1 = 0; \pi_2 = \pi_3 = \pi_4 = 2$.

Recompute state of each arc.

Iteration 2

Arc (i, j)	π_i	$-\pi_j$	$-b_{ij}$	q_{ij}	f_{ij}	State	In kilter?
(1, 2)	0	−2	2	0	$0 = l_{12}$	B	Yes
(1, 3)	0	−2	5	3	$2 > l_{13}$	A_2	No
(2, 3)	2	−2	1	1	$0 = l_{23}$	A	Yes
(2, 4)	2	−2	3	3	$2 > l_{24}$	A_2	No
(3, 2)	2	−2	1	1	$2 > l_{32}$	A_2	No
(3, 4)	2	−2	6	6	$0 < l_{34}$	A_1	No
(4, 1)	2	0	0	2	$2 < l_{41}$	A_1	No

1. Arc (1, 2) is still in kilter but its state has changed from A to B; therefore, flow can be increased in (1, 2) *without* driving it out of kilter.
2. State of arc (1, 3) is A_2; decrease f_{13} to l_{13}.
3. Find path from node 1 to node 3 by the labeling procedure.

Labeling procedure		
Node	Label	
1	$(3^-, 2)$	[Decrease flow in arc (1, 3)]
2	$(1^+, 2)$	[Increase flow in arc (1, 2)]
3	$(2^-, 2)$	[Decrease flow in arc (3, 2)]

Breakthrough has occurred.
Change flow in the cycle as indicated by labels.
Recompute state of each arc.

Iteration 3

Arc (i, j)	π_i	$-\pi_j$	$-b_{ij}$	q_{ij}	f_{ij}	State	In kilter?
(1, 2)	0	−2	2	0	$2 = u_{12}$	B	Yes
(1, 3)	0	−2	5	3	$0 = l_{13}$	A	Yes
(2, 3)	2	−2	1	1	$0 = l_{23}$	A	Yes
(2, 4)	2	−2	3	3	$2 > l_{24}$	A_2	No
(3, 2)	2	−2	1	1	$0 = l_{32}$	A	Yes
(3, 4)	2	−2	6	6	$0 < l_{34}$	A_1	No
(4, 1)	2	0	0	2	$2 < l_{41}$	A_1	No

1. Pick the next out-of-kilter arc (2, 4).
2. State of arc (2, 4) is A_2; decrease f_{24} to l_{24}.
3. Find path from node 2 to node 4 by labeling procedure.

Labeling procedure	
Node	Label
2	$(4^-, 2)$
1	$(2^-, 2)$
3	Cannot be labeled; (1, 3) is in kilter
4	Cannot be labeled; flow decrease drives (4, 1) more out of kilter

Nonbreakthrough has occurred.

$X = \{1, 2\}$, $\bar{X} = \{3, 4\}$, $M = \{(1, 3), (2, 3), (2, 4)\}$, $\bar{M} = \Phi$, $D = \min(3, 1, 3) = 1$, $\bar{D}$ = does not exist, $D_f = 1$.

New π values: $\pi_1 = 0$, $\pi_2 = 2$, $\pi_3 = \pi_4 = 3$.

Recompute state of each arc.

Iteration 4

Arc (i, j)	π_i	$-\pi_j$	$-b_{ij}$	q_{ij}	f_{ij}	State	In kilter?
(1, 2)	0	−2	2	0	$2 = u_{12}$	B	Yes
(1, 3)	0	−3	5	2	$0 = l_{13}$	A	Yes
(2, 3)	2	−3	1	0	$0 = l_{23}$	B	Yes
(2, 4)	2	−3	3	2	$2 > l_{24}$	A_2	No
(3, 2)	3	−2	1	2	$0 = l_{32}$	A	Yes
(3, 4)	3	−3	6	6	$0 < l_{34}$	A_1	No
(4, 1)	3	0	0	3	$2 < l_{41}$	A_1	No

1. Arc (2, 3) is still in kilter, but its state has changed from A to B; therefore, flow can be increased in (2, 3) *without* driving the arc out of kilter.
2. State of arc (2, 4) is A_2; decrease f_{24} to l_{24}.
3. Find path from node 2 to node 4 by labeling procedure.

Labeling procedure	
Node	Label
2	$(4^-, 2)$
3	$(2^+, 1)$
4	$(3^+, 1)$

Breakthrough has occurred.
Change flow in the cycle as indicated by labels.
Recompute state of each arc.

Iteration 5

Arc (i, j)	π_i	$-\pi_j$	$-b_{ij}$	q_{ij}	f_{ij}	State	In kilter?
(1, 2)	0	-2	2	0	$2 = u_{12}$	B	Yes
(1, 3)	0	-3	5	2	$0 = l_{13}$	A	Yes
(2, 3)	2	-3	1	0	$1 > l_{23}$	B	Yes
(2, 4)	2	-3	3	2	$1 > l_{24}$	A_2	No
(3, 2)	3	-2	1	2	$0 = l_{32}$	A	Yes
(3, 4)	3	-3	6	6	$1 = l_{13}$	A	Yes
(4, 1)	3	0	0	0	$2 < l_{41}$	A_1	No

1. Arc (3, 4) has been brought into kilter.
2. Arc (2, 4) is still out of kilter.
3. State of arc is A_2 ; reduce f_{24} to l_{24} .
4. Find path from node 2 to node 4 by labeling procedure.

Labeling procedure

Node	Label
2	$(4^-, 1)$
1	$(2^-, 1)$
3	Cannot be labeled; (1, 3) is in kilter
4	Cannot be labeled; a flow decrease drives (4, 1) more out of kilter

Nonbreakthrough has occurred.

$X = \{1, 2\}$, $\bar{X} = \{3, 4\}$, $M = \{(1, 3), (2, 4)\}$, $\bar{M} = \Phi$, $D = \min(2, 2) = 2$, $\bar{D}$ = does not exist, $D_f = 2$.

New π values: $\pi_1 = 0$, $\pi_2 = 2$, $\pi_3 = \pi_4 = 5$.

Recompute state of each arc.

Iteration 6

Arc (i, j)	π_i	$-\pi_j$	$-b_{ij}$	q_{ij}	f_{ij}	State	In kilter?
(1, 2)	0	−2	2	0	$2 = u_{12}$	B	Yes
(1, 3)	0	−5	5	0	$0 = l_{13}$	B	Yes
(2, 3)	2	−5	1	−2	$1 = u_{23}$	C	Yes
(2, 4)	2	−5	3	0	$1 > l_{24}$	B	Yes
(3, 2)	5	−2	1	4	$0 = l_{32}$	A	Yes
(3, 4)	5	−5	6	6	$1 = l_{13}$	A	Yes
(4, 1)	5	−0	0	5	$2 < l_{41}$	A_1	No

1. Arc (2, 4) has been brought into kilter.
2. Arc (4, 1) is out of kilter.
3. State of the arc is A_1; increase f_{41} to l_{41}.
4. Find path from arc 1 to arc 4 by labeling procedure.

Labeling procedure	
Node	Label
1	$(4^+, 1)$
2	Cannot be labeled: arc (1, 2) is in kilter
3	$(1^+, 1)$
4	Cannot be labeled: arc (3, 4) is in kilter

Nonbreakthrough has occurred.

$X = (1, 3)$, $\bar{X} = (2, 4)$, $M = \{(3, 2), (3, 4)\}$, $\bar{M} = (2, 3)$, $D = \min(4, 6) = 4$, $\bar{D} = \min(2) = 2$, $D_f = 2$.

New π values: $\pi_1 = 0$, $\pi_2 = 4$, $\pi_3 = 5$, $\pi_4 = 7$.

Recompute state of each arc.

Iteration 7

Arc (i, j)	π_i	$-\pi_j$	$-b_{ij}$	q_{ij}	f_{ij}	State	In kilter?
(1, 2)	0	−4	2	−2	$2 = u_{12}$	C	Yes
(1, 3)	0	−5	5	0	$0 = l_{13}$	B	Yes
(2, 3)	4	−5	1	0	$1 = u_{13}$	B	Yes
(2, 4)	4	−7	3	0	$1 > l_{24}$	B	Yes
(3, 2)	5	−4	1	2	$0 = l_{32}$	A	Yes
(3, 4)	5	−7	6	4	$1 = l_{34}$	A	Yes
(4, 1)	7	−0	0	7	$2 < l_{41}$	A_1	No

1. Arc (2, 3) has changed from state C to state B; therefore flow can be decreased in (2, 3) without driving it out of kilter.
2. State of arc (4, 1) is A_1; increase f_{41} to l_{41}.
3. Find path from node 1 to node 4 by labeling procedure.

Labeling procedure	
Node	Label
1	$(4^+, 1)$
3	$(1^+, 1)$
2	$(3^-, 1)$
4	$(2^+, 1)$

Breakthrough has occurred.

In fact, arc (4, 1) is brought into kilter; therefore we have found the optimum solution. The following tabulation gives the optimum flow values for all the arcs.

Arc (i, j)	π_i	$-\pi_j$	$-b_{ij}$	q_{ij}	f_{ij}	State	In kilter?
(1, 2)	0	−4	2	−2	$2 = u_{12}$	C	Yes
(1, 3)	0	−5	5	0	$l_{12} \leq 1 \leq u_{12}$	B	Yes
(2, 3)	4	−5	1	0	$0 = l_{23}$	B	Yes
(2, 4)	4	−7	3	0	$l_{24} \leq 2 \leq u_{24}$	B	Yes
(3, 2)	5	−4	1	2	$0 = l_{32}$	A	Yes
(3, 4)	5	−7	6	4	$1 = l_{34}$	A	Yes
(4, 1)	7	−4	0	7	$3 = l_{41}$	A	Yes

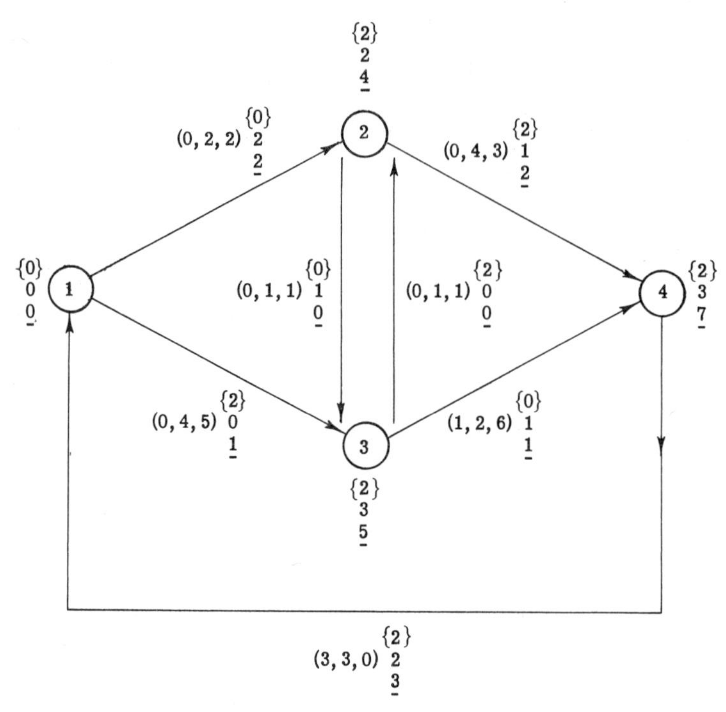

Fig. 3.13 A minimum-cost circulation example: flow values and node numbers in iterations, 2, 5, and 7. For each arc the ordered triple is $(l_{ij}, u_{ij}, -b_{ij})$; the flow is $\{f_{ij}\}$ in iteration 2, f_{ij} in iteration 5, and $\underline{f_{ij}}$ in iteration 7 (the optimal solution). The corresponding node numbers are $\{\pi_i\}$, π_i, and $\underline{\pi_i}$.

the progress toward the optimal solution after iterations 2, 5, and 7, respectively.

3.4. Application of the Out-of-Kilter Algorithm to the Operational Policy Problem

As indicated previously, all the constraints for the OKA must be linear. In the mathematical model of the OP problem [expressions (2.8)–(2.17)] all the constraints except (2.10) are linear. Constraint (2.10) was linearized as follows:

$$\sum_j K_j \sum_m A_{jm} Q_{imt} \geq P_{it} \qquad \text{for all } i \text{ and } t$$

where K_j is the amount of energy produced by turbine j per acre-foot of water.

If it is assumed that each turbine operates at a constant head and at a constant speed, there is a linear relationship between flow and energy.

In the example water resources problem (fully described in Chapter 4) the OKA is repeatedly applied to solve the OP problem. Sections 4.2 and 4.3 describe how the OKA optimizes the flow variables of the original system network whose configuration is detailed in Tables B.3 and B.4. The solution procedure followed is identical to that described in Sections 3.3.1–3.3.7 and utilized in Section 3.3.8. However, the complexity of the problem precludes the listing of intermediate results.

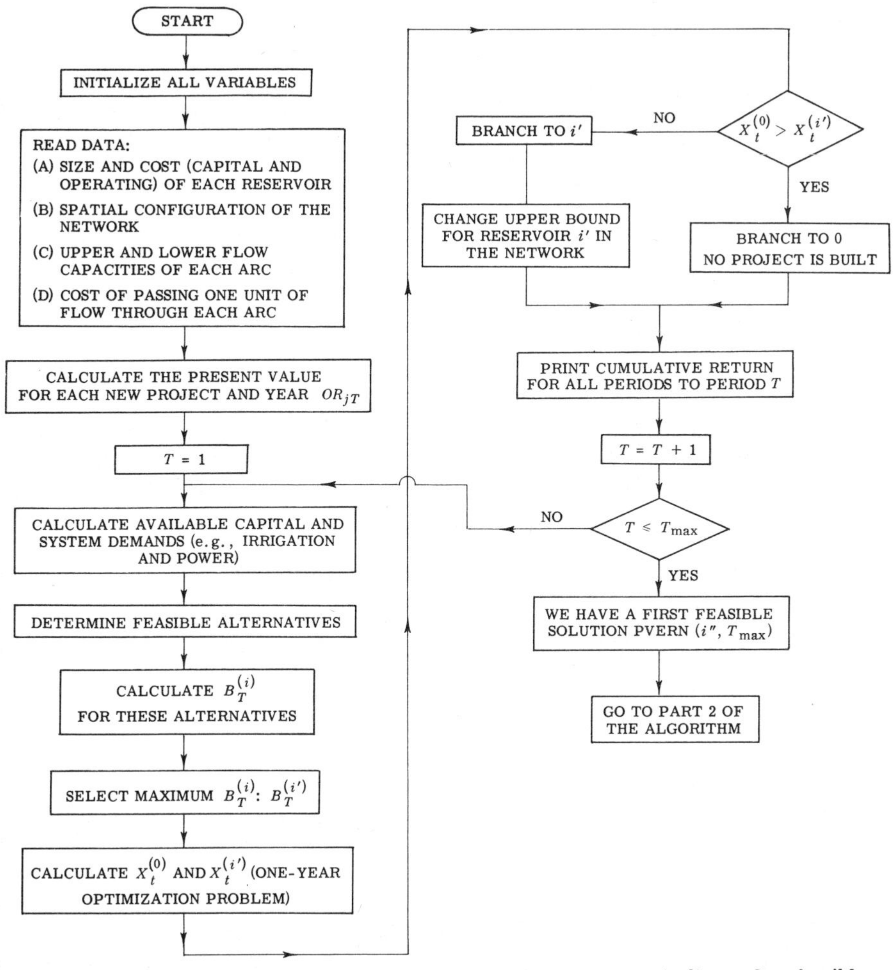

Fig. 3.14 Flow chart of the optimization algorithm (part 1): finding a first feasible solution.

3.5. Flow Chart of the Algorithm for Solving the Water Resources Problem

Figures 3.14 and 3.15 describe every major step in the combined branch and bound and out-of-kilter algorithm. Figure 3.14 shows how the capital

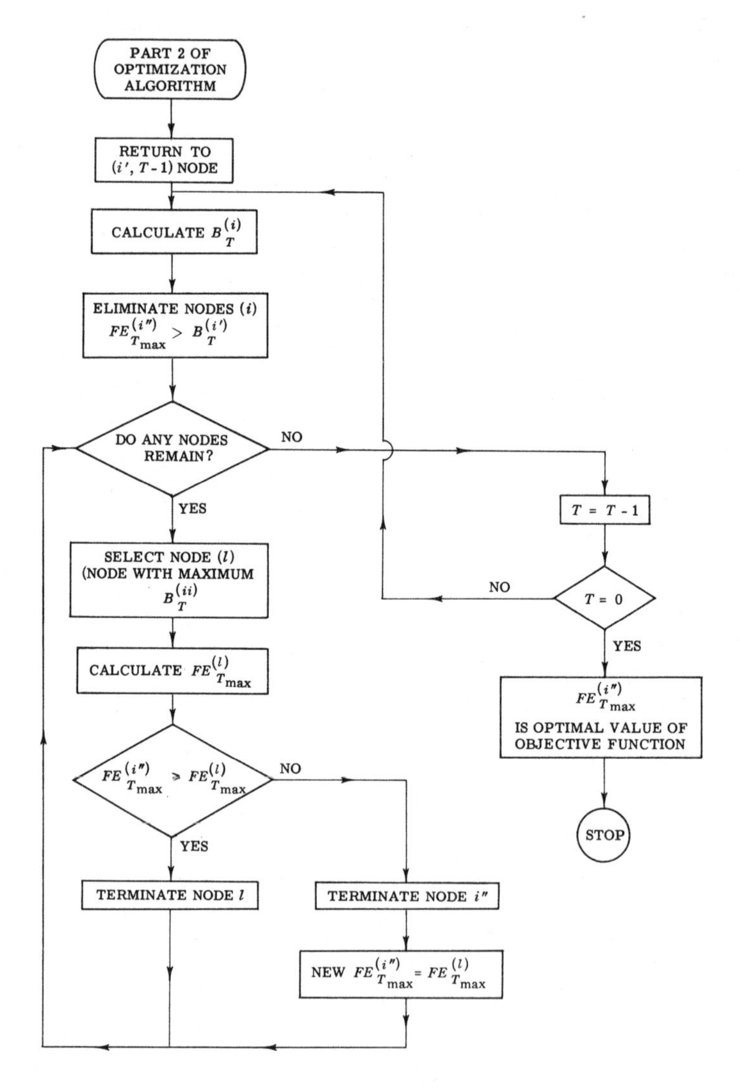

Fig. 3.15 Flow chart of the optimization algorithm (part 2): the backtracking sequence.

budgeting problem is connected with the operating policy problem (a one-year optimization problem) and how the first feasible solution $FE_{j''t}$ is found. Figure 3.15 shows the backtracking sequence in which each node is examined and eliminated or accepted as the new feasible solution. Appendix A gives the complete documentation for the algorithm together with detailed flow charts and a listing of the FORTRAN IV computer program.

3.6. Summary

In this chapter we have explored the various optimization techniques that could be used to resolve the problem formulated in Chapter 2. Some possible solution techniques (the generalized Lagrange multiplier technique, dynamic programming, and mixed integer programming) were discussed and rejected because of the characteristics of the optimal expansion problem or inherent intractabilities of the technique itself. The algorithm finally chosen takes advantage of the discreteness of the variables in the original problem (Problem I) (1) to decompose it into the set of all feasible combinations (termed Problem II) and (2) to consider the economic return for each combination (termed Problem III). The combination of (1) and (2) with the best return is necessarily the optimum solution for Problem I.

Problem II is a capital budgeting (CB) problem and is solved using Little's branch and bound algorithm (BBA). Problem III is called the operating policy (OP) problem and is solved by Fulkerson's out-of-kilter algorithm (OKA) which takes advantage of the problem network structure. The two problems are interconnected because the operating return in the CB problem is determined only by obtaining the *optimal solution* of the OP problem, and the total number of dams in the OP problem is the *optimal solution* of the CB problem.

This chapter provides the theoretical background of Little's and Fulkerson's algorithms, mentions some of their general applications, and demonstrates their methodology with simple and detailed examples. Finally the application of the BBA and OKA algorithms to the optimal expansion problem is explained.

Chapter 4

APPLICATION OF THE OPTIMIZATION ALGORITHM TO A WATER RESOURCES SYSTEM

4.1. The Water Resources System

The water resources system whose configuration is shown in Fig. 4.1 will be used as an illustration of the application of the algorithm described in Chapter 3. The configuration represents the energy and irrigation needs of a hypothetical river basin [hereafter called the model for development (MD) river basin] and is similar to the existing configuration in the Maule River Basin, which lies in Central Chile between latitudes 35°10″ S and 36°20″ S [McLaughlin, 1967; Wallace, 1966].

The MD river basin lies in a semiarid region that has an annual rainfall of about 29 in. The bulk of the rainfall (about 60%) occurs in the winter. In the MD river basin approximately 1.5 million acres are irrigated, and half of the irrigation waters are surface waters. Five hundred forty million kilowatt-hours of hydroelectric energy are generated annually. Because of projected increasing industrialization and depletion of ground-water resources, demands for hydroelectric energy and irrigation surface water will increase. To meet the prospective demands on the surface-water resources, new impoundments are proposed for construction at technologically feasible sites in the river basin.

Four dams (represented by nodes 1, 4, 22, 28) are used for flow regulation. Ten unregulated streams (represented by arcs $U_1, U_2, \ldots, U_{10}$) are the physical inputs to the system. The specified schedule of increasing energy and irrigation needs of the river basin can be met by building a suitable combination of dams. Three possible dam sites exist (represented by nodes 14, 21, and 27) and any of three sizes of dam may be built at each site. A planning horizon of 50 yr has been chosen, and the problem is

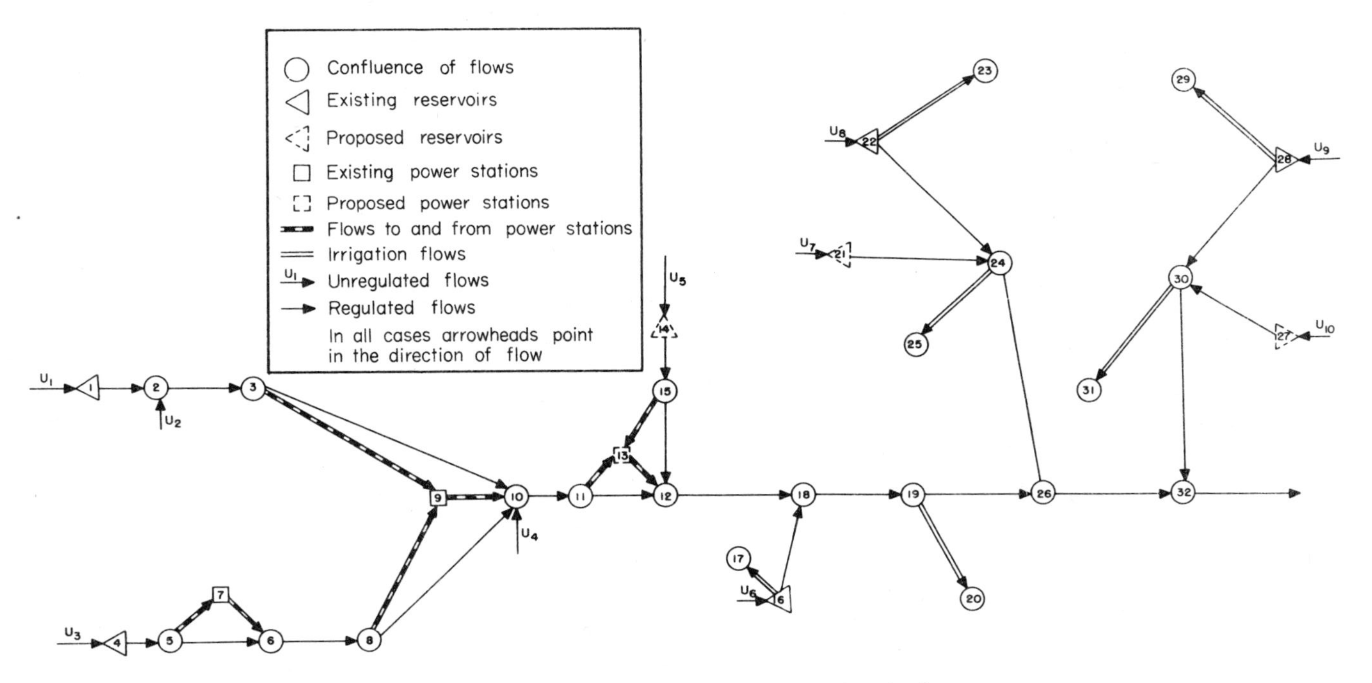

Confluence of flows
Existing reservoirs
Proposed reservoirs
Existing power stations
Proposed power stations
Flows to and from power stations
Irrigation flows
Unregulated flows
Regulated flows
In all cases arrowheads point in the direction of flow

Fig. 4.1 Spatial configuration of the MD river basin.

when and where to locate each new dam. Table 4.1 shows the location and active storage capacities of old and new dams and the economic data for the new dams. Observe that reservoir 6, 7, or 8 must be built at location 14; reservoir 9, 10, or 11 must be built at location 21; reservoir 12, 13, or 14 must be built at location 27.

The maximum annual return MR_i for reservoir i is the net revenue obtained by operating the reservoir at its optimum level. For example, the annual return MR_6 for the hydroelectric reservoir 6 is calculated as follows.

The turbines associated with reservoir 6 can process 300,000 acre-ft of water in each period, with a net return of \$4/acre-ft in the summer period and \$5/acre-ft in the winter. Thus

$$MR_6 = 4(300{,}000) + 5(300{,}000) = \$2.7 \text{ million}$$

The maximum return MR_{12} for the irrigation reservoir 12 is found as follows. The active storage of reservoir 12 is 200,000 acre-ft. If the reservoir is full at the beginning of the irrigation season (the summer period), all of the water may be used for irrigation *plus* the unregulated flows in the

Table 4.1

Some Economic and Physical Data on Reservoirs

Reservoir number	Represented by node	Active storage capacity (10^4 acre-ft)	Required investment (10^6 \$)	Maximum annual return (10^6 \$)	Useful economic life (yr)
Existing reservoirs					
1	1	40			
2	4	30			
3	16	18			
4	22	30			
5	28	40			
New reservoirs					
6	14	60	40.0	2.7	50
7	14	50	38.0	2.43	50
8	14	40	34.0	2.16	50
9	21	12	6.3	0.36	50
10	21	10	5.5	0.30	50
11	21	8	4.55	0.24	50
12	27	20	10.0	0.60	50
13	27	15	8.4	0.48	50
14	27	10	5.55	0.36	50

irrigated season. The net revenue per acre-foot of supplied water is \$3. Thus

$$MR_{12} = 3(200{,}000) = \$600{,}000$$

The available capital at the beginning of the planning period $\hat{C}_0$ was \$2.8 million.

It was decided to limit the objectives of the proposed expansion to two disparate facets of a water resources system: (1) a withdrawal–consumptive use and (2) a withdrawal–nonconsumptive use. A withdrawal–consumptive use of water implies that water is withdrawn from the system, is consumed, and is *not* returned to the system. A withdrawal–nonconsumptive use of water implies that it is withdrawn and later returned to the system, usually downstream from the withdrawal point. Irrigation was chosen as the withdrawal consumptive use of water, since it is more consumptive than other uses; hydroelectric energy was chosen as the withdrawal nonconsumptive use because it is a function of power head and water volume.

The other common purposes of water resources development, namely municipal, industrial, and recreational use, are not incorporated *explicitly* in this example. Municipal and industrial demands are satisfied only through the specification of minimum flows in the adjacent river reaches, while recreational needs are covered by the specification of minimum seasonal water levels in the reservoirs.

4.2. Network Configuration

In the mathematical model, the year is divided into two periods (winter and summer). Table 4.2 lists the upper (high $\equiv C_m$) and lower (low $\equiv L_m$) flow capacities of each arc in the system and also the cost of passing one unit of water through each arc (cost coefficient) of the system, as it exists at the beginning of period 1. Note that all of these coefficients are integer values because the OKA requires that all variables and costs be integer. Also, the node numbers of Fig. 4.1 refer to the summer period. The winter node numbers are obtained by adding 40 to every summer node number, e.g., node 50 of the winter period corresponds to node 10 of the summer period.

The unit irrigation and energy revenues (arcs 41–54 and 95–102) obtainable from the use of 1 acre-ft of water are similar to those applicable in the Maule River Basin in Chile and in the Lewiston region of Idaho [Bower, 1962; Wallace, 1966]. However, they are given negative values to

Table 4.2a
Network Data for the Two-Period System—Time Period Number 1, Summer

Arc number	Nodes $(i\text{–}j)$	Cost coefficient[a] c_{ij}	Limits[a] High	Limits[a] Low	Comments
1	239–1[b]	0	40	40	Unregulated flows
2	239–2	0	15	15	
3	239–4	0	30	30	
4	239–10	0	10	10	
5	239–14	0	70	70	
6	239–16	0	12	12	
7	239–21	0	11	11	
8	239–22	0	20	20	
9	239–27	0	15	15	
10	239–28	0	30	30	
11	239–1	0	20	20	Reservoir initial conditions
12	239–4	0	15	15	
13	239–14	0	0	0	
14	239–16	0	9	9	
15	239–21	0	0	0	
16	239–22	0	30	30	
17	239–27	0	0	0	
18	239–28	0	40	40	
19	1–2	0	80	0	Regulated flows
20	2–3	0	90	0	
21	3–10	0	70	0	
22	4–5	0	60	0	
23	5–6	0	40	0	
24	6–8	0	70	0	
25	8–10	0	60	0	
26	10–11	0	190	0	
27	11–12	0	190	0	

Arc number	Nodes $(i\text{–}j)$	Cost coefficient c_{ij}	Limits High	Limits Low	Comments
28	14–15	0	80	0	Regulated flows
29	15–12	0	80	0	
30	12–18	0	290	0	
31	16–18	0	25	0	
32	18–19	0	300	0	
33	19–26	0	300	0	
34	22–24	0	70	0	
35	21–24	0	50	0	
36	24–26	0	80	8	
37	26–32	0	330	0	
38	28–30	0	60	0	
39	27–30	0	50	0	
40	30–32	0	95	8	
41	5–7	0	40	0	Energy flows
42	7–6	−3	15	0	
43	3–9	0	55	0	
44	8–9	0	40	0	
45	9–10	−2	20	0	
46	11–13	0	250	0	
47	13–12	−4	0	0	
48	15–13	0	60	0	
49	16–17	−3	20	0	Irrigation flows
50	19–20	−2	40	0	
51	22–23	−4	30	0	
52	24–25	−3	15	0	
53	28–29	−4	30	0	
54	30–31	−3	35	0	

[a] All upper and lower flow limits are in 10^4 acre-ft. All costs are in $/acre-ft.

[b] 239–1 means flow goes from node 239 to node 1.

Table 4.2b
Network Data for the Two-Period System—Time Period Number 2, Winter

Arc number	Nodes (i–j)	Cost coefficient[a] c_{ij}	Limits[a] High	Limits[a] Low	Comments	Arc number	Nodes (i–j)	Cost coefficient c_{ij}	Limits High	Limits Low	Comments
55	1–41[b]	0	20	20	Reservoir initial conditions	82	54–55	0	80	0	Regulated flows
56	4–44	0	15	15		83	55–52	0	80	0	
57	14–54	0	0	0		84	52–58	0	290	0	
58	16–56	0	18	0		85	56–58	0	25	0	
59	21–61	0	0	0		86	58–59	0	300	0	
60	22–62	0	30	8		87	59–66	0	300	0	
61	27–67	0	0	0		88	62–64	0	70	0	
62	28–68	0	40	12		89	61–64	0	70	0	
63	239–41	0	60	60	Unregulated flows	90	64–66	0	80	8	
64	239–42	0	20	20		91	66–72	0	300	0	
65	239–44	0	50	50		92	68–70	0	60	0	
66	239–50	0	30	30		93	67–70	0	50	0	
67	239–54	0	60	60		94	70–72	0	95	8	
68	239–56	0	25	25		95	45–47	0	40	0	Energy flows
69	239–61	0	15	15		96	47–46	−4	15	0	
70	239–62	0	40	40		97	43–49	0	55	0	
71	239–67	0	30	30		98	48–49	0	40	0	
72	239–68	0	60	60		99	49–50	−3	20	0	
73	41–42	0	80	0	Regulated flows	100	51–53	0	250	0	
74	42–43	0	90	0		101	53–52	−5	0	0	
75	43–50	0	70	0		102	55–53	0	60	0	
76	44–45	0	80	0		103	56–57	0	0	0	Irrigation flows
77	45–46	0	60	0		104	59–60	0	0	0	
78	46–48	0	70	0		105	62–63	0	0	0	
79	48–50	0	60	0		106	64–65	0	0	0	
80	50–51	0	190	0		107	68–69	0	0	0	
81	51–52	0	190	0		108	70–71	0	0	0	

[a] All upper and lower flow limits are in 10^4 acre-ft. All costs are in $/acre-ft.

[b] 1–41 means flow goes from node 1 to node 41.

Table 4.2c

Network Data for the Two-Period System—Exit Flows

Arc number	Nodes (i–j)	Cost coefficient[a] c_{ij}	Limits[a] High	Limits[a] Low	Comments
109	41–240[b]	0	20	20	"Final" reservoir conditions
110	44–240	0	15	15	
111	54–240	0	0	0	
112	56–240	0	9	9	
113	61–240	0	0	0	
114	62–240	0	30	30	
115	67–240	0	0	0	
116	68–240	0	40	40	
117	32–240	0	400	0	Exit flow (Period 1)
118	72–240	0	400	0	Exit flow (Period 2)
119	17–240	0	20	0	Exit irrigation flows
120	20–240	0	40	0	
121	23–240	0	30	0	
122	25–240	0	90,000	0	
123	29–240	0	50	0	
124	31–240	0	90,000	0	
125	57–240	0	20	0	
126	60–240	0	30	0	
127	63–240	0	40	0	
128	65–240	0	150	0	
129	69–240	0	50	0	
130	71–240	0	350	0	
131	240–239	0	90,000	0	

[a] All upper and lower flow limits are in 10^4 acre-ft. All costs are in \$/acre-ft.
[b] As in other tables.

transform the OP problem from a minimum-cost circulation problem to one of circulation at maximum revenue. It should be noted that no irrigation revenue is obtainable in the winter season and also that energy revenues are higher in winter than in summer.

Nodes 239 and 240 are the source and sink nodes, respectively. Note that the unregulated flow inputs to the system (arcs 1–10 and 63–72) and the original reservoir levels (arcs 11–18) originate from the source node

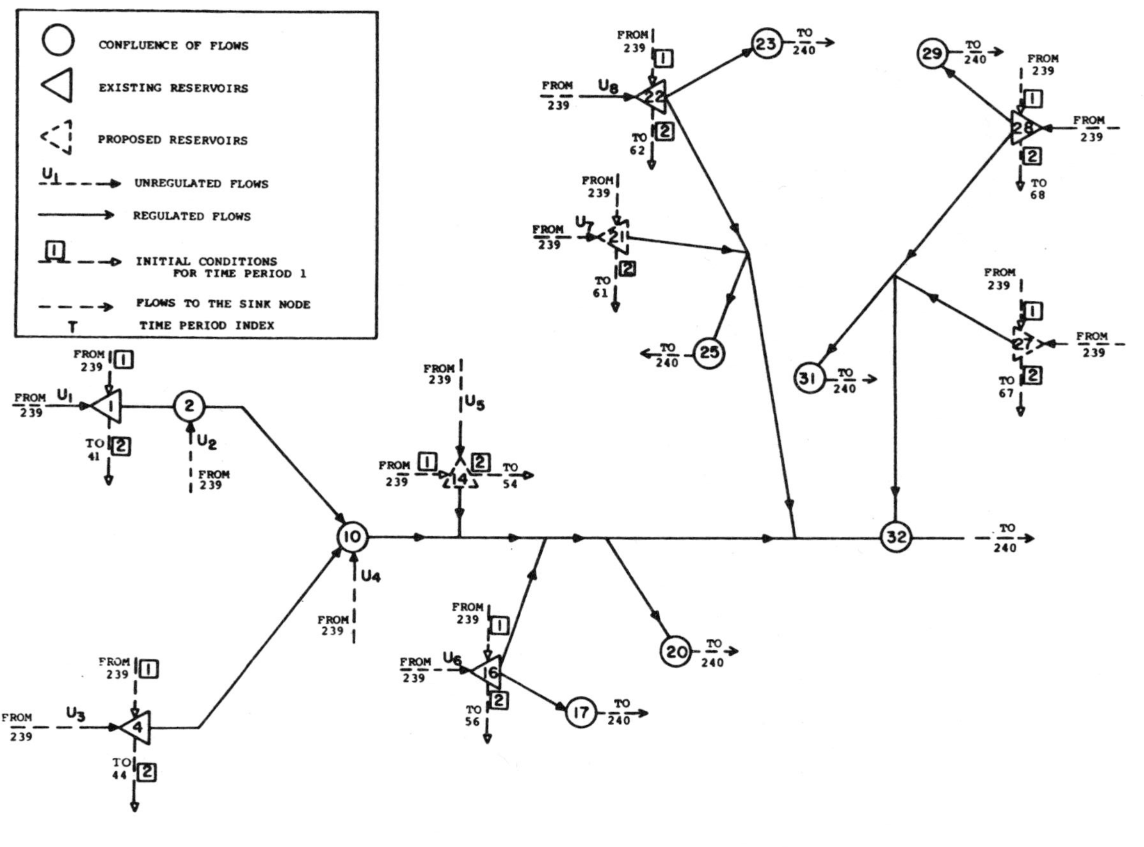

Fig. 4.2 Information flow diagram for inputs and outputs of the MD river basin (time period 1).

(239). The flows in arcs 1–10, 63–72, and 11–18 are assured by setting the upper and lower bounds of these arcs at the same value. The consumptively used irrigation waters are removed from the system to the sink node (arcs 119–130).

Municipal and industrial demands have been taken into account by specifying the minimum flows in arcs 36 and 40 of period 1 (summer) and in arcs 90 and 94 of period 2 (winter). Recreation demands are satisfied by specifying the minimum flows in arcs 60, 62, 114, and 116. All exit flows from the system (arcs 117 and 118) and all final reservoir volumes (arcs 109–116) are removed to the sink node. An artificial arc (131) connects the sink node (240) to the source node (239) to complete the circulation network. Figure 4.2 shows the information flow for inputs and outputs to the system.

Note that the OKA solves the linear programming problem known as the minimum-cost circulation problem. This entails that all inputs to the system (incoming hydrological flows, reservoir initial conditions) must come from the source node. Likewise all outputs from the system (consumptively used irrigation waters, outgoing hydrological flows, reservoir final conditions) must go to the sink node.

Table 4.3 shows the changes in the arc capacities that will occur if any of the proposed new dams are added to the system. Figure 4.3 shows the profile of future irrigation demands that will occur in arcs 52 and 54, respectively.

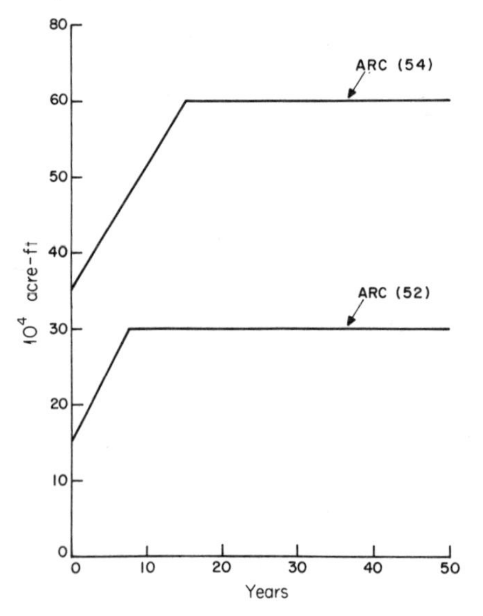

Fig. 4.3 Demand schedule for irrigation.

Table 4.3

Network Parameter Changes When New Reservoirs Are Introduced

Reservoir number	Arc number	Nodes $(i–j)$	Cost coefficient c_{ij}	Limits[a] High	Limits[a] Low	Reservoir number	Arc number	Nodes $(i–j)$	Cost coefficient c_{ij}	Limits High	Limits Low
6	13	239–14	0	30	30	10	15	239–21	0	10	10
6	47	13–12	−4	30	0	10	59	21–61	0	10	0
6	57	14–54	0	60	0	10	113	61–240	0	10	10
6	101	53–52	−5	30	0						
6	111	54–240	0	30	30	11	15	239–21	0	8	8
						11	59	21–61	0	8	0
7	13	239–14	0	25	25	11	113	61–240	0	8	8
7	47	13–12	−4	27	0						
7	57	14–54	0	50	0	12	17	239–27	0	20	20
7	101	53–52	−5	27	0	12	61	27–67	0	20	0
7	111	54–240	0	25	25	12	115	67–240	0	20	20
8	13	239–14	0	20	20	13	17	239–27	0	15	15
8	47	13–12	−4	24	0	13	61	27–67	0	15	0
8	57	14–54	0	40	0	13	115	67–240	0	15	15
8	101	53–52	−5	24	0						
8	111	54–240	0	20	20	14	17	239–27	0	10	10
						14	61	27–67	0	10	0
9	15	239–21	0	12	12	14	115	67–240	0	10	10
9	59	21–61	0	12	0						
9	113	61–240	0	12	12						

[a] All upper and lower flow limits are in 10^4 acre-ft.

4.3. Solving the Operational Policy Problem

In this section we discuss how the OKA optimizes the flow variables of the original system network whose configuration is detailed in Tables B.3 and B.4. The listing in Appendix B includes the source and sink nodes, and the values of the cost and the upper and lower flow limits corresponding to each of the 131 arcs of the network. Twelve network arcs had (negative) nonzero costs (called NZC arcs); data on the NZC arcs are detailed in Table 4.5a.

The objective function used had the form

Minimize

$$\hat{X}_t = -(b_{7,6}f_{7,6} + b_{9,10}f_{9,10} + \cdots + b_{53,52}f_{53,52})$$

Zero initial values were assumed for all the node (π_i) numbers and a feasible nonoptimal flow circulation was imposed on the network. The solution procedure followed was identical to that described in Sections 3.3.1–3.3.7 and utilized in Section 3.3.8. However, the complexity of the problem precludes the listing of intermediate results. The optimal solution is listed in Tables B.8–B.10; a summary of the NZC arcs is in Table 4.5a and 4.5b. Note that the node numbers in the optimal solution remain unchanged (consult Table B.10). Consequently, for the arcs with zero costs (called ZC arcs), the optimal q_{ij} values are zero; thus any feasible flow in the ZC arcs *may* be the optimal solution. On the other hand, the values of q_{ij} for the NZC arcs remain negative, and the optimal flows in these arcs must equal their respective upper bounds. It is explicitly assumed that imported water can be brought to all the dams of the system. If perchance there are some dams to which imported water cannot be brought, a value of ∞ can be used for the appropriate canal cost.

The only information that was prescribed for reservoir operating rules pertained to the hydroelectric reservoirs. Constant-head hydroelectric generators were used in the model, and to ensure a constant head, an initial condition was forced for each reservoir generating electrical energy for each period. For example, referring to Fig. 4.1, reservoir 4 is used for generating electric power through arc (5–7). Prescribing similar values for the high and low limits in the arcs listed in the following tabulation ensured that there was always exactly this amount of water in the reservoir, so that the constant-head turbines could be operated. Excluding the restrictions of the constant head, the reservoir operating rules are determined by the OKA.

Arc number	Nodes	Cost coefficient	Limits		Comments
			High	Low	
12	239–4	0	15	15	Initial condition for period 1
56	4–44	0	15	15	Initial condition for period 2
110	44–240	0	15	15	Final condition for period 2

Indirectly the hydrology affects the optimal answer insofar as the OKA distributes optimally the *supplied* water through the network.

The notation used in the statement of the OP problem is not completely compatible with that of the OKA used in Sections 3.3.1–3.3.5 for two reasons:

1. Once the network has been set up, the OKA does not distinguish between water flows that are used for different purposes (e.g., energy generation and irrigation); on the contrary, it uses a common coefficient and flow notation for *all* arcs. (Refer to Table 3.1.)
2. Similarly, the network notation does not differentiate between the different time periods. The record of inputs and coefficients for the different time periods is maintained by the bookkeeping of the user of the OKA.

4.4. Solving the Capital Budgeting Problem

For these reasons all network flows and coefficients have been designated by the notation of Table 3.1. In this notation an example of the capital budgeting problem objective function, Eq. (2.1), would be as follows. (Only the nonzero cost coefficients are included; the nodes are designated by the subscripts on the b's.)

Maximize

$$\sum_{t=1}^{T_{\max}} \alpha(b_{7,6}f_{7,6} + b_{9,10}f_{9,10} + \cdots + b_{53,52}f_{53,52})$$

net operating return from the present set of subsystems over the planning period

$$+ \sum_{t=1}^{T_{max}} \alpha (b_{13,12} f_{13,12} + b_{53,52} f_{53,52})$$

net operating return from the newly added subsystems

$$- \sum_{t=1}^{T_{max}} \alpha \times 10^6 [\lambda_{6,t}(40.0) + \lambda_{7,t}(6.3) + \cdots + \lambda_{14,t}(5.55)]$$

capital cost of the projects over the planning period

The λ values are determined by the optimization algorithm. Table 4.5a lists the cost coefficients b_{ij} for the first term, Table 4.3 lists those for the second term, and Table 4.1 lists those for the third term. Arcs with nonzero cost coefficients will be referred to as NZC arcs.

We next describe in detail how the bound $B_{7,2}$ is calculated for the example problem; refer to Section 3.2.4 and Fig. 4.4 and Table 4.4. Here $B_{7,2}$ is the economic return associated with the introduction of reservoir 7 in year 2. It is found by applying equation (3.4) as follows:

$$B_{7,2} = CR_1 + TR_2 + OR_{j2} + OR_{j'3} + OR_{j''4} + \cdots + OR_{j^m 50}$$

Since projects 7, 8, 9 are mutually exclusive, only six projects are considered in year 3, i.e., reservoirs 9 to 14 (inclusive). Reservoir 12 is chosen because it has the largest value of OR_{jt}. Projects 12 to 14 are mutually exclusive, and so only projects 9 to 11 are considered in year 4. Reservoir 9 has the largest value of OR_{jt} and so was picked. In years 5 through 50 there were no remaining projects to be built; hence the equation for $B_{7,2}$ becomes

$$B_{7,2} = CR_1 + TR_2 + OR_{7,2} + OR_{12,3} + OR_{9,4}$$

Because no new projects were built in year 1, Eqs. (3.1) and (3.2) were applied to find CR_1 and TR_2 as follows:

$$CR_1 = \sum_{t=1}^{1} [\hat{X}_1/(1+r)^t]$$

$$TR_2 = \hat{X}_2 \sum_{t=2}^{T_{max}} [1/(1+r)^t]$$

As a further consequence of building no projects in year 1, $\hat{X}_2 = \hat{X}_1$. Therefore

$$CR_1 + TR_2 = \hat{X}_1 \sum_{t=1}^{T_{max}} [1/(1+r)^t]$$

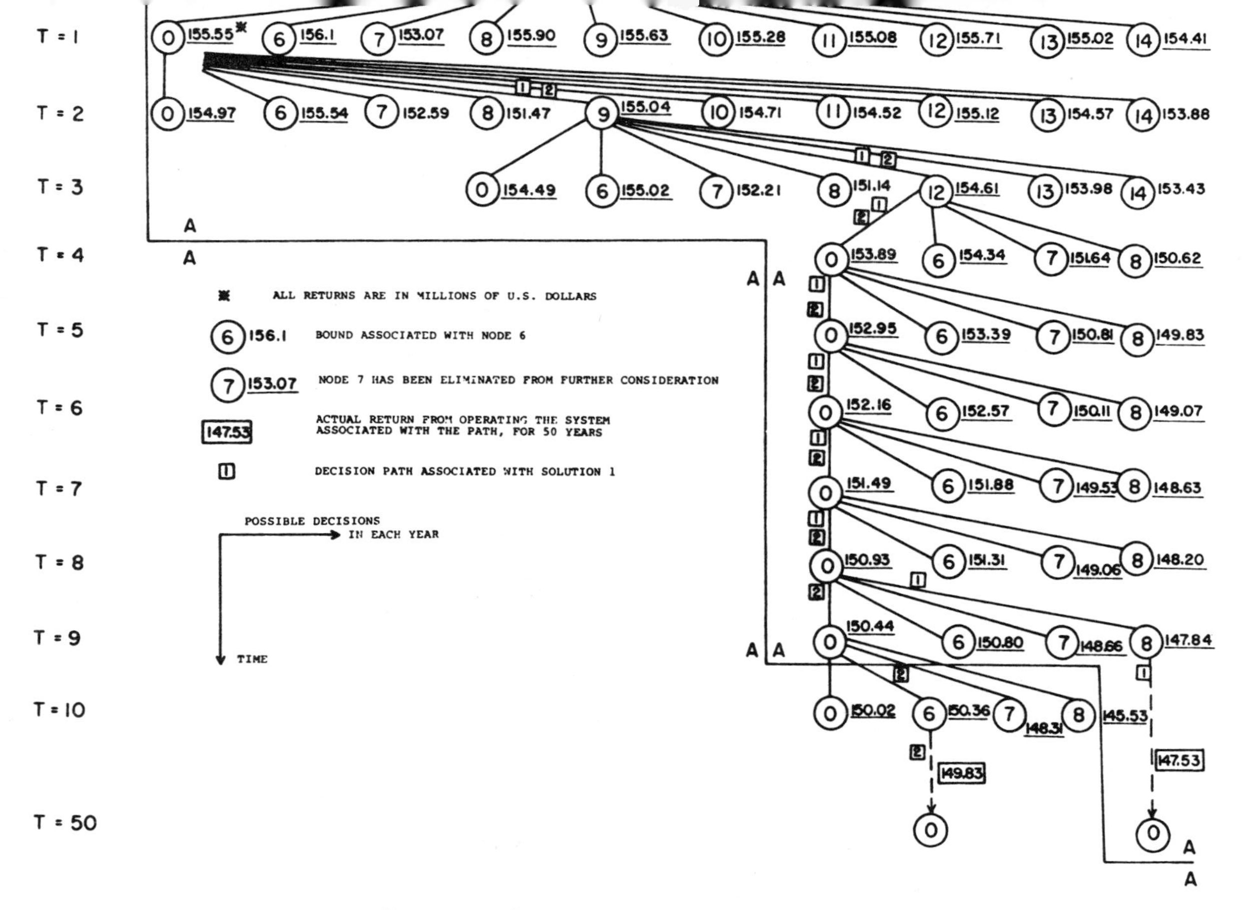

Fig. 4.4 Decision tree for the first two solutions.

Table 4.4

The Bound Matrix[a]

Year	New reservoirs									No reservoir
	6	7	8	9	10	11	12	13	14	0
1	0	0	0	0	0	0	0	0	0	0
2	0	152599208	151476433	0	154712578	154518900	0	154456746	153875713	154965415
3	0	152208459	151135317	0	0	0	0	153983884	153428536	154493764
4	0	0	0	0	0	0	0	0	0	0
5	0	0	0	0	0	0	0	0	0	0
6	0	0	0	0	0	0	0	0	0	0
7	0	0	0	0	0	0	0	0	0	0
8	0	0	0	0	0	0	0	0	0	0
9	0	0	0	0	0	0	0	0	0	0
10	0	0	0	0	0	0	0	0	0	150444348
11	0	0	0	0	0	0	0	0	0	0
12	0	0	0	0	0	0	0	0	0	0
13	0	0	0	0	0	0	0	0	0	0
14	0	0	0	0	0	0	0	0	0	0
15	0	0	0	0	0	0	0	0	0	0
16	0	0	0	0	0	0	0	0	0	0
17	0	0	0	0	0	0	0	0	0	0
18	0	0	0	0	0	0	0	0	0	0
19	0	0	0	0	0	0	0	0	0	0
20	0	0	0	0	0	0	0	0	0	0
21	0	0	0	0	0	0	0	0	0	0
22	0	0	0	0	0	0	0	0	0	0
23	0	0	0	0	0	0	0	0	0	0
24	0	0	0	0	0	0	0	0	0	0

25	0	0	0	0	0	0	0	0	0	0
26	0	0	0	0	0	0	0	0	0	0
27	0	0	0	0	0	0	0	0	0	0
28	0	0	0	0	0	0	0	0	0	0
29	0	0	0	0	0	0	0	0	0	0
30	0	0	0	0	0	0	0	0	0	0
31	0	0	0	0	0	0	0	0	0	0
32	0	0	0	0	0	0	0	0	0	0
33	0	0	0	0	0	0	0	0	0	0
34	0	0	0	0	0	0	0	0	0	0
35	0	0	0	0	0	0	0	0	0	0
36	0	0	0	0	0	0	0	0	0	0
37	0	0	0	0	0	0	0	0	0	0
38	0	0	0	0	0	0	0	0	0	0
39	0	0	0	0	0	0	0	0	0	0
40	0	0	0	0	0	0	0	0	0	0
41	0	0	0	0	0	0	0	0	0	0
42	0	0	0	0	0	0	0	0	0	0
43	0	0	0	0	0	0	0	0	0	0
44	0	0	0	0	0	0	0	0	0	0
45	0	0	0	0	0	0	0	0	0	0
46	0	0	0	0	0	0	0	0	0	0
47	0	0	0	0	0	0	0	0	0	0
48	0	0	0	0	0	0	0	0	0	0
49	0	0	0	0	0	0	0	0	0	0
50	0	0	0	0	0	0	0	0	0	0

[a] A nonzero entry indicates the value of a bound for the corresponding node. A zero entry indicates that the node has been eliminated from further consideration.

In this case

$$\hat{X}_1 = \$7{,}350{,}000, \quad T_{\max} = 50 \text{ yr}, \quad r = 4.625\%$$

Upon substituting these values one obtains

$$CR_1 + TR_2 = \$142{,}346{,}000$$

The values for the OR terms are found in the PVBAR matrix in Appendix B:

$$OR_{7,2} = \$8{,}278{,}000, \quad OR_{12,3} = \$1{,}415{,}000, \quad OR_{9,4} = \$561{,}000$$

Therefore

$$B_{7,2} = 142{,}346{,}000 + 8{,}278{,}000 + 1{,}415{,}000 + 561{,}000 = \$152{,}600{,}000$$

In Figure 4.4 the value of $B_{7,2}$ is truncated to the first 5 significant figures of \$152.60 million.

4.5. Results of the Optimization

The detailed results obtained from the overall optimization algorithm are listed in Appendix B. It took 6.138 sec on a CDC 6600 to generate a first solution and 297.12 sec to backtrack completely through the decision tree. In the backtracking procedure, 124 alternative solutions were examined; of these, eleven had a higher return than the first solution. Only those alternative solutions whose bounds *at the time of examination* were higher than the best solution were considered. The efficacy of the heuristic procedure employed in finding a first feasible solution to the optimization problem is demonstrated by the fact that the return from the first solution was within 3.3% of the final optimum.

Irrigation demands were never fully met in any year in the optimal solution because there were insufficient surface waters available to meet the simultaneous depletion of ground-water resources and the overall increase in irrigation acreage. Note that the first solution scheduled first the construction of the irrigation reservoirs 9 and 12 and the hydroelectric dam 8, while the optimal solution reversed this order and also scheduled the construction of the first dam at a later date.

The situation of constructing the irrigation reservoirs first would be achieved by using a penalty function related to the failure to meet irrigation demands. Another advantage would also accrue from using a penalty function: many solutions whose bounds heretofore were higher than the best feasible solution would not be examined, in view of their having lower bounds than the best feasible solution. As a result, the total computation time would be reduced significantly. In Chapter 5 the efficacy of using a penalty function is discussed in greater detail.

4.5.1. Comparison of Operating Policies for Initial and Final Configurations

The differences in operating policies for different system configurations were elucidated by studying the NZC arcs. Tables 4.5a, 4.5b, 4.6a, and 4.6b show the effects of the operating policies on the flows and dual variables associated with each NZC arc for the configuration that existed initially and the final optimal configuration.‡

For the initial configuration (i.e., before any new dams are added) all flows in the NZC arcs equaled their corresponding upper-limit constraints. Furthermore, the values of the dual γ_{ij} variables corresponded to the absolute value of the cost coefficient for each respective arc and thereby indicated that the resources of the initial configuration were being stretched to the limit. For example, the flows in arcs 42, 45, 49 through 54, 96, and 99 equaled the upper limit for each respective arc, so that the *installed* electrical and irrigation facilities (including pumps, canals, and so on) were being used *constantly* at maximum capacity.

The final network configuration differs from the original configuration in a number of respects:

1. The upper bounds of arcs 47 and 101 were raised from 0 to 30, indicating that hydroelectric facilities were added.
2. The flows in both arcs 47 and 101 increased from 0 to 30 and the flows in arcs 52 and 54 increased from 15 to 27 and 35 to 55, respectively; the overall supply of hydroelectric energy and irrigation water was augmented.
3. Four of the γ_{ij} variables were not equal to the absolute value of the corresponding cost coefficient. The γ_{ij} variables of arcs 52 and 54 were equal to zero because the flows in these arcs are less than their respective upper bounds. No revenue improvement would result from raising the upper bounds on flows for these arcs.

Arcs 51 and 53 had values of γ_{ij} less than the absolute value of the corresponding cost coefficient. Examination of the configuration around arc

‡ The dual problem associated with the operational policy problem and the dual variables are discussed and defined in Section 3.3.

Table 4.5a

Data for Nonzero-Cost Arcs in the Initial Configuration

Arc number	Nodes (i–j)	Cost coefficient[a] c_{ij}	Limits[a] High	Limits[a] Low	Flow[b]	Comments	
42	7–6	−3	15	0	15	Energy	Summer
45	9–10	−2	20	0	20	Energy	Summer
47	13–12	−4	0	0	0	Energy	Summer
49	16–17	−3	20	0	20	Irrigation	Summer
50	19–20	−2	40	0	40	Irrigation	Summer
51	22–23	−4	30	0	30	Irrigation	Summer
52	24–25	−3	15	0	15	Irrigation	Summer
53	28–29	−4	30	0	30	Irrigation	Summer
54	30–31	−3	35	0	35	Irrigation	Summer
96	47–46	−4	15	0	15	Energy	Winter
99	49–50	−3	20	0	20	Energy	Winter
101	53–52	−5	0	0	0	Energy	Winter

[a] All upper and lower flow limits are in 10^4 acre-ft. All costs are in \$/acre-ft.
[b] Flow is the *actual* flow in each arc, as defined by the OKA.

Table 4.5b

Data for Dual Variables Associated with Nonzero-Cost Arcs in the Initial Configuration

Arc number	Source node i	Sink node j	π_i	π_j	c_{ij}	q_{ij}	γ_{ij}[a]	δ_{ij}
42	7	6	0	0	−3	−3	3	0
45	9	10	0	0	−2	−2	2	0
47	13	12	0	0	−4	−4	4	0
49	16	17	0	0	−3	−3	3	0
50	19	20	0	0	−2	−2	2	0
51	22	23	0	0	−4	−4	4	0
52	24	25	0	0	−3	−3	3	0
53	28	29	0	0	−4	−4	4	0
54	30	31	0	0	−3	−3	3	0
96	47	46	0	0	−4	−4	4	0
99	49	50	0	0	−3	−3	3	0
101	53	52	0	0	−5	−5	5	0

[a] Increasing the upper bound of each arc by one unit will increase the total revenue by the corresponding γ_{ij} value.

Table 4.6a

Data for Nonzero-Cost Arcs in the Final Configuration

Arc number	Nodes $(i\text{–}j)$	Cost coefficient[a] c_{ij}	Limits[a] High	Low	Flow[b]	Comments
42	7–6	−3	15	0	15	Summer
45	9–10	−2	20	0	20	
47	13–12	−4	30	0	30	
49	16–17	−3	20	0	20	
50	19–20	−2	40	0	40	
51	22–23	−4	30	0	30	
52	24–25	−3	30	0	27	
53	28–29	−4	30	0	30	
54	30–31	−3	65	0	55	
96	47–46	−4	15	0	15	Winter
99	49–50	−3	20	0	20	
101	53–52	−5	30	0	30	

[a] All upper and lower flow limits are in 10^4 acre-ft. All costs are in \$/acre-ft.
[b] Flow is the *actual* flow in each arc as defined by the OKA.

Table 4.6b

Data for Dual Variables Associated with Nonzero-Cost Arcs in the Final Configuration

Arc number	Source node i	Sink node j	π_i	π_j	c_{ij}	q_{ij}	γ_{ij}[a]	δ_{ij}
42	7	6	0	0	−3	−3	3	0
45	9	10	0	0	−2	−2	2	0
47	13	12	0	0	−4	−4	4	0
49	16	17	0	0	−3	−3	3	0
50	19	20	0	0	−2	−2	2	0
51	22	23	3	0	−4	−1	1	0
52	24	25	3	0	−3	0	0	0
53	28	29	3	0	−4	−1	1	0
54	30	31	3	0	−3	0	0	0
96	47	46	0	0	−4	−4	4	0
99	49	50	0	0	−3	−3	3	0
101	53	52	0	0	−5	−5	5	0

[a] Increasing the upper bound of each arc by one will increase the total revenue by the corresponding γ_{ij} value.

51 shows the reason for this discrepancy, namely that arc 51 *competes* for available water with arc 52. If the upper bound of arc 51 were raised by one unit, an extra unit of flow would result in the arc because a higher return per unit flow results in arc 51 than in arc 52. Consequently, the flows in arc 34 (from node 22 to node 24) and in arc 52 were reduced by one unit.

net effect of raising bound 51 by one unit	=	benefits gained by *increasing* flow by one unit in arc 51	−	benefits lost by *decreasing* flow by one unit in arc 52

$$= 4 - 3 = 1$$

Similar reasoning is used to explain the discrepancy between γ_{ij} and the cost coefficient for arc 54.

4.5.2. Comments on Some Feasible Solutions

Figure 4.4 shows the paths through the solution tree (composed of nodes and arcs) of the first feasible solution and of one other solution that had a higher return than the first solution. The solution tree may also be interpreted as a matrix (called BOUND) as shown for the first feasible solution in Table 4.4, and examination of matrix BOUND (also see Appendix B) at any time tells which possible solutions have been examined and which remain to be examined.

Matrix BOUND in the example problem has 10 × 50 elements. The columns represent the nine possible dams that may be constructed and the decision to undertake no project. The rows represent the planning horizon (50 yr). The nonzero elements in the matrix represent those nodes that remain to be examined. The zero elements represent those nodes that have been eliminated from further consideration for one of the following reasons: (1) they have been examined already; (2) the bound associated with the node(s) is lower than the current feasible solution; (3) the dam associated with the node is infeasible for that year because of the capital budget limit.

The elements of matrix BOUND correspond to the nodes of the solution tree. Since column 1 corresponds to dam 6, column 2 to dam 7, and so on, the matrix elements for columns 1 through 9 have been modified so that they are identified by the dam numbers. For example, element (2,7) in the matrix in the revised nomenclature is known as (7, 7), since column 2 represents dam 7. Column 10 corresponds to the decision to build no project, and the elements in this column are known by the no-project number (0). The correspondence of the elements of the matrix to the nodes of the solution tree may be demonstrated by comparing the BOUND matrix that occurs just after the first feasible solution as tabulated in the computer print-out in Table 4.4 with Fig. 4.4. For example, the value of element (7, 2) of matrix BOUND is equal to 152.60 million, and this corresponds exactly to the value of the bound of node (7, 2), i.e., dam 7 in year 2, of Table 4.4. Similarly, the value of element (0, 9) of the matrix, 150.44, corresponds exactly to the value of the bound of node (0, 9).

The nodes of the tree to the right of line A–A of Fig. 4.4 correspond to the nodes represented by the elements of the bound B_{it} as it is constituted after the first solution has been found. The algorithm then backtracks until it finds a node with a bound higher than the current feasible solution [node (0, 9) with a bound of 150.44 million]. Branching from node (0, 9) gives a solution with a higher return (solution 2) than the first feasible solution. The algorithm then backtracks to the node with the highest bound (0, 3) greater than the solution *at the time of scrutiny* (solution 2) and branches forward (in time) through the tree to discover whether the return from the solution associated with node (0, 3) is in fact higher than solution 2. The procedure of backtracking and tracking forward is repeated until all the elements of matrix BOUND are zero.

4.6. Estimation of Computer Execution Time to Calculate the First Feasible Solution

It would be highly desirable to be able to predict the total computation time of the overall optimization algorithm for a water resources management problem, given the number of possible projects and the size of the network configuration. Unfortunately, this is impossible because the total number of solutions that must be examined cannot be predicted, a priori. However, it is possible to forecast the calculation time needed to obtain the first solution, because the number of computation steps is predictable. The algorithm calls mainly on two subroutines (OLERSEN and BOUND) to obtain the first solution, and knowing the total computing time for these two subroutines gives a good estimate of the time to calculate first feasible solution.

The main program (program DAMBLD) calls subroutine OLERSEN (which in turns calls subroutine NETFLO) at least once for every year and at the most twice. It calls OLERSEN twice in one year when the possibility of a new dam being added in that year is under consideration and once when no reservoir will be built in that year. The time needed for running OLERSEN is very short because the initial conditions of the decision variables of the subroutine are close to the optimal values. Subroutine

BOUND is called once by the main program for each year. The running time for BOUND is greatest in the earlier years (when the number of possible choices is greatest) and then decreases accordingly as reservoirs are built, thereby reducing the number of choices for succeeding years.

In the example described above, a library subroutine of the CDC 6600 computation system (subroutine SECOND) was used to isolate the calculation time for subroutines OLERSEN and BOUND. For 50 years of calculations, subroutine OLERSEN was called 53 times for a total computing time of 3.735 sec; individual computing times ranged from 0.112 to 0.044 sec. Subroutine BOUND was called 50 times for a total computing time of 2.099 sec; individual computing times ranged from 0.238 to 0.019 sec. The combined computing time for all calculations in subroutines OLERSEN and BOUND was 5.824 sec, which was approximately 95% of the total computing time needed to obtain a first feasible solution (6.138 sec).

For a problem having a different network configuration, the following steps are needed to estimate the computation time needed to find a first feasible solution:

1. Use routine OLERSEN as a main program and empirically find its running time (t_{OLER}) with the problem network configuration as data. Use different sets of perturbed data to get an average running time. Estimate the number of times OLERSEN will be called by the main program (N_{OLER}).
2. Determine the running time for BOUND ($t_{i\mathrm{BND}}$) for different combinations of possible reservoir choices. Estimate the number of years each combination of projects is under consideration ($Y_{i\mathrm{BND}}$).
3. The calculation time for obtaining a first feasible solution equals

$$(1.10)\ [(t_{\mathrm{OLER}})(N_{\mathrm{OLER}}) + \sum_i (t_{i\mathrm{BND}})(Y_{i\mathrm{BND}})]$$

where the factor 1.10 allows for the fact that some of the computation takes place outside of these subroutines.

From experience in working with problems of a given configuration, one can estimate in advance the total number of solutions that need to be examined before the optimum solution is found. If S is the total number of such solutions, then the total calculation time for obtaining the optimum solution equals

$$(1.10)\ S[(t_{\mathrm{OLER}})(N_{\mathrm{OLER}}) + \sum_i (t_{i\mathrm{BND}})(Y_{i\mathrm{BND}})]$$

The algorithm provides a monotonically increasing range of solutions, from the first feasible solution to the optimal solution. Thus if the user of the algorithm felt at any time that the running time for the program was becoming excessive, he could set a computation time limit and still obtain a feasible solution—one that was close to the optimum.

4.7. Summary

In this chapter, we have used the expansion of the water resources system of a hypothetical river basin (called the MD river basin) to illustrate the efficacy of the decomposition algorithm proposed in Chapter 3. The objectives of the proposed expansion were limited to two disparate facets of a water resources system: (1) withdrawal consumptive use (irrigation) and (2) withdrawal nonconsumptive use (generation of hydroelectric energy). Other common purposes of water resources development, namely, municipal, industrial, and recreational use, were incorporated *implicitly* in the river basin model by specifying respectively minimum flows in the relevant river reaches and minimum seasonal water levels in the reservoirs. The model incorporated realistic economic and hydrological parameters.

The OKA solved the operating policy problem in a straightforward fashion. Reservoir operating rules were not prespecified except for constant-head hydroelectric reservoirs. In this case identical high and low limits were prespecified in the relevant arcs to maintain a constant head in the reservoirs. No special problems were encountered with Little's BBA.

The detailed results found are listed in Appendix B. The objective functions used and an example of a bound calculation are exhibited in Sections 4.3 and 4.4. Highlights from the output data are tabulated, illustrated, and discussed in Section 4.5. Of special interest is the role of the γ_{ij} dual variables in spotlighting those parts of the river basin where the water supply does not meet all the demands. The sensitivity of decisions to parameter changes and the use of penalty functions is discussed in Chapter 5.

The efficacy of the solution procedure was demonstrated by the fact that the return from the first feasible solution was within 3.3% of the final optimum. In Section 4.6 a heuristic method is explained for calculating the computer execution time to reach a feasible solution of any similar problem.

Chapter 5

THE SENSITIVITY OF PLANNING DECISIONS IN RIVER BASIN MANAGEMENT

In the previous chapters the emphasis has been placed on finding an optimal plan for water resources development. However, other aspects of the optimal plan besides the values of the objective function and the variables must be taken into account by planners. For example, all plans must be executed in a dynamic environment in which changes in inputs, costs, and objectives occur because of related technological, political, and social changes. If a planner is to select an appropriate series of decisions in time, he must be aware of the consequences of such changes on the optimal plan(s). In fact, it often is more important for the planner to know how the physical system will perform under different operating conditions than it is for him to devise an optimal plan for a single postulated set of prices, inputs, and demands.

Methods of incorporating the appropriate features of variability and fluctuations into the planning include deterministic techniques, such as "worst case" methods and sensitivity analysis, as well as stochastic approaches. To accommodate the prospects of uncertainty, in this chapter we first examine the sensitivity of planning decisions by application of the algorithm previously used in Chapter 4 to solve the deterministic river basin management problem. We then make a few remarks concerning the stochastic approach to water resources planning, even though this topic is essentially outside the scope of this book.

5.1. Sensitivity Analysis

A sensitivity analysis examines the effect of small perturbations in model inputs and model parameters on the values of the model outputs and the revenue function. For example, a 10% change in the value of a model input or parameter could have a negligible or significant effect on a constraint or on the return, depending on whether the variable of interest is insensitive or sensitive to changes in the model input or parameter. Figure 5.1 illustrates how the total system cost (not discounted), i.e., the capital cost plus operation and maintenance cost, changes as the projected total requirements for water vary above or below the projected base requirement in the Texas water development plan [Texas Water Development Board, 1971]. A 10% increase in the water requirements from the base case (250 million acre-ft) increases the total cost from $9.2 to 10.3 billion, or approximately 12%. From the opposite viewpoint, Fig. 5.2 shows what percentage change is required in certain inputs and demands to produce a 10% change in the total system cost. The length of each bar in Fig. 5.2 is inversely proportional to the importance of the parameter to the system cost response. If the outputs change very little with perturbations in the parameters and the inputs, then the coefficients and the inputs in the equations for these system model may be quite approximate. On the other hand, if one response shows a high degree of sensitivity to changes in its

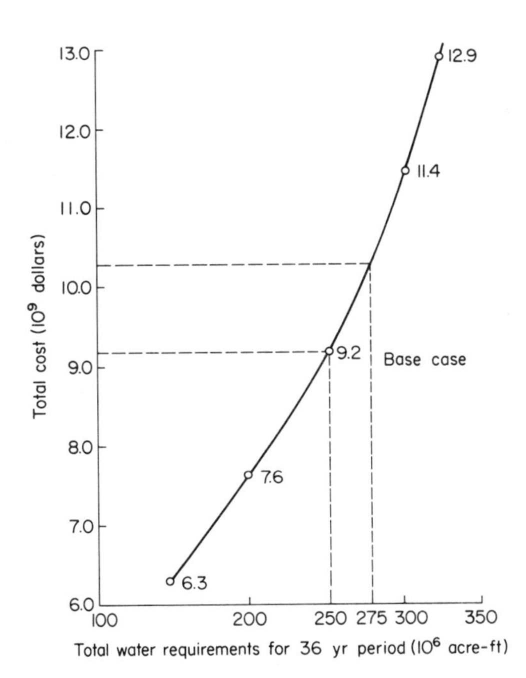

Fig. 5.1 Variation of cost of total system with changes in water requirements.

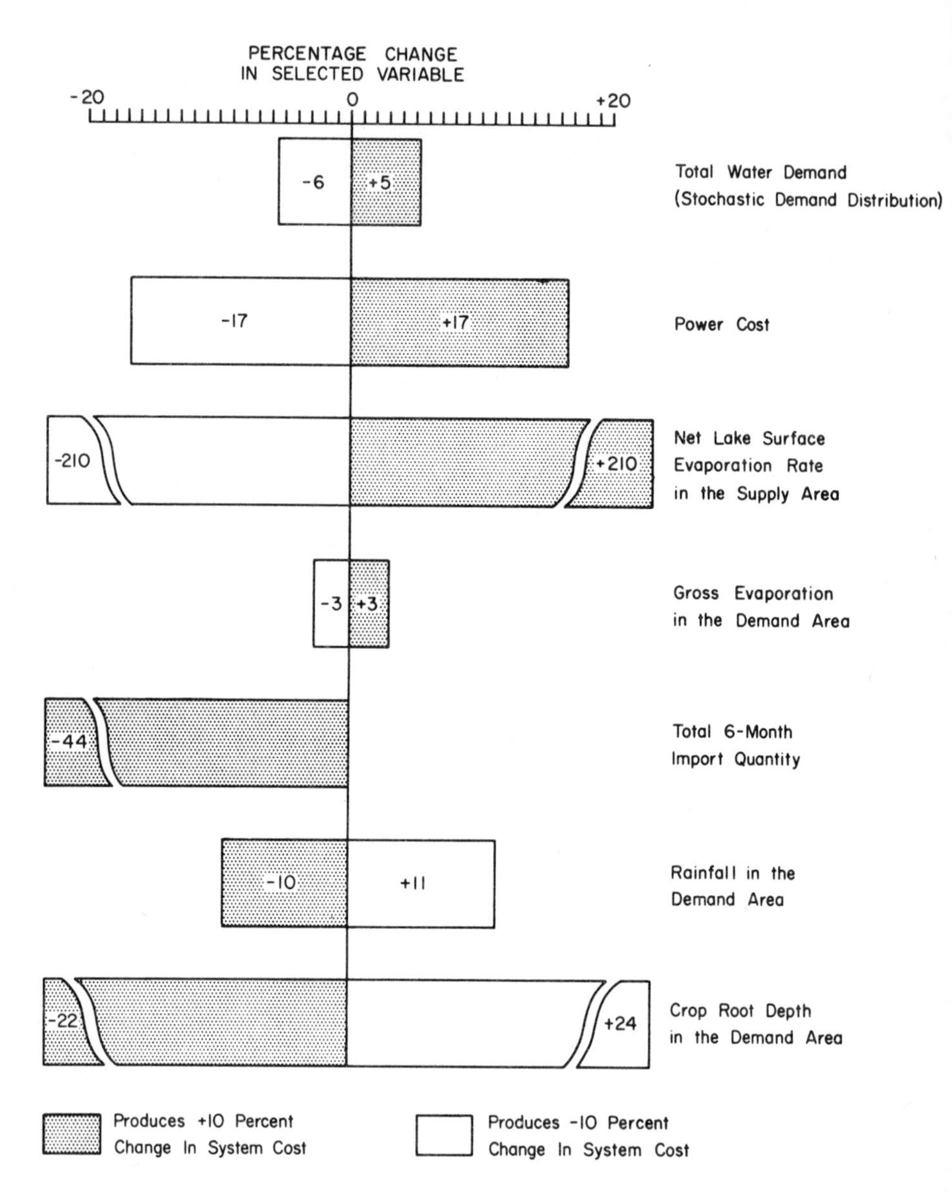

Fig. 5.2 Percentage changes in selected variables that produce a corresponding 10% change in total system cost. The shaded areas represent a 10% increase in cost while the unshaded areas represent a 10% decrease in cost. The shorter the length of the bar the greater the influence of the variable on total system cost.

parameters and inputs, the coefficients and the inputs for the response must be determined with much greater accuracy. A sensitivity analysis of a river basin model thus will reveal how accurate the parameters and inputs must be. There is no point in meticulously obtaining data for relationships that can be shown to play little part in the final analysis, and vice versa.

If a river basin is composed of a number of subsystems, each representing a somewhat arbitrarily defined region or activity in the total river basin, it is possible to examine the effect of input and parameter perturbations both on the subsystem outputs and on the total system outputs. Because each equation or relationship in a subsystem model also contains one or more parameters or coefficients, it is also desirable, in the search for appropriate parameter values, to be able to evaluate how sensitive portions of the system may be to uncertainties in the parameters or disturbances in the inputs.

To sum up, a sensitivity analysis on a river basin model can be used to (1) prevent placing undue emphasis on any portion of the real basin, either in the planning or execution stage; (2) help determine the scope of the model of the system in light of the existing or future criteria for evaluation; and (3) aid in developing and comparing alternate courses of action. Thus a sensitivity analysis can make the planning for a regional river basin more economic in terms of time, money, and effort.

5.2. The Sensitivity Function

In this section we define single- and multiparameter sensitivity functions. Keep in mind that analytical evaluation of the functions is not particularly easy, but, on the other hand, that numerical evaluation of the functions is quite time consuming. If a function f depends on a single parameter x, we write the function as $f = f(x)$ to abbreviate the notation, and define the sensitivity of f with respect to x as

$$S_x^f = (df/f)/(dx/x) = (d \ln f)/(d \ln x) \tag{5.1}$$

The first ratio indicates that the sensitivity represents the fractional change in f divided by the fractional change in x, while the second ratio indicates that we are concerned with relative changes rather than absolute changes in defining the sensitivity. The sensitivity is evaluated at nominal or design values of x, and for reasonably small changes in x the fractional change in $f(x)$ is

$$df/f = S_x^f \, dx/x \tag{5.2}$$

Suppose now that the function f depends on n parameters

$$\mathbf{x} = [x_1, x_2, \ldots, x_n]^T,$$

i.e., $f = f(\mathbf{x})$. To obtain a multiparameter sensitivity, we make use of the first-order terms of a Taylor series or the total differential. Using the latter we obtain

$$df = \frac{\partial f}{\partial x_1} dx_1 + \cdots + \frac{\partial f}{\partial x_n} dx_n$$

which can be rearranged as follows:

$$\frac{df}{f} = \left[\frac{x_1}{f}\frac{\partial f}{\partial x_1}\right]\frac{dx_1}{x_1} + \cdots + \left[\frac{x_n}{f}\frac{\partial f}{\partial x_n}\right]\frac{dx_n}{x_n} \tag{5.3}$$

By analogy with Eq. (5.2) the terms in square brackets represent the sensitivities of $f(\mathbf{x})$ to each of the parameters x_i, that is,

$$S_{x_i}^f = \frac{x_i}{f}\frac{\partial f}{\partial x_i} \tag{5.4}$$

Consequently we can write

$$df/f = S_{x_1}^f (dx_1/x_1) + \cdots + S_{x_n}^f (dx_n/x_n) \tag{5.5}$$

Equation (5.5) relates the fractional change in $f(\mathbf{x})$ to the fractional change in each of the parameters; f can represent the objective function value, dependent variable values, constraint values, and so forth.

If the dependence of f on $\mathbf{x}$ is a known unconstrained function and the partial derivatives of $f(\mathbf{x})$ can be computed analytically (or numerically), the evaluation of (df/f) is reasonably straightforward. Once the sensitivities are determined from Eq. (5.4), it is clear from Eq. (5.5) which of the x_i are the most important in determining the overall sensitivity (df/f). Equation (5.5) can be used to find the change in $f(\mathbf{x})$ from any base case for any change in the parameters and can also be used for a "worst case" analysis as well.

For system models with constraints, if the functions and equations in the model are linear, the sensitivity analysis may be carried out in a very systematic fashion. Special variations of the simplex method of linear programming, such as the revised simplex method and parametric linear programming, combined with duality analysis have been used to obtain the sensitivity coefficients (dual variables) for oil refinery operations [Wilde and Beightler, 1967, Chapter 2] and water resources systems [Wallace, 1966, McLaughlin, 1967].

However, for systems that are described by combinatorial mathematical models, such as the water resources system analyzed in Chapters 2–4, computation of the sensitivity coefficients is considerably more complicated. The nonconvexity of the set of feasible solutions of combinatorial models precludes automatic computation of the sensitivity coefficients from the optimal solution. For combinatorial models the number of possible sets of perturbed inputs is myriad, and an enormous amount of computer time would be needed to find the sensitivity effects of each set. Consequently, it is opportune to limit a sensitivity analysis to discovering the effects of what are believed to be the most significant inputs on the values of the objective function.

5.3. Relationship between Outputs and Inputs of the Water Resources Model

The relationship between all the inputs and outputs of the mathematical model of the water resources system that was developed in Chapter 2 may be represented conceptually by a matrix as follows:

	Outputs							
Inputs	f,	λ_{jt},	D_{ijt},	...	X_{ijt},	value of constraint (2.2),	...,	value of constraint (2.17)
H_{ijt}	x							
C_{jt}	x							
⋮								
I_{ijt}	x				x			
⋮								
U_{ijt}								
V_j								x
β_{jt}						x		
⋮								
Q_{imt}					x			

The inputs to the model include all the constants and the decision variables. The outputs of the system include the return function f, all the state variables, and the values of the inequalities and equations in the set (2.2)–(2.17). The effects of the inputs on the values of the constraints are needed to establish whether or not a constraint is violated. An x in element (k, l)

of the matrix signifies that output l is affected by input k. No entry signifies no relationship between l and k. For example, consider the dependent variables that determine the year of construction of each dam, λ_{jt}. Note that

$$\boldsymbol{\lambda}_j = \begin{bmatrix} \lambda_{j1} \\ \lambda_{j2} \\ \vdots \\ \lambda_{jt} \end{bmatrix}$$

is a vector that shows the year in which project j is constructed and the pattern of investment involved, e.g., the vector

$$\boldsymbol{\lambda}_3 = [0, 1, 1, 1, 0, 0, \ldots, 0]^T$$

signifies that project 3 was built in year 2 and the financing of the project was spread over 3 years. The sensitivity coefficients show how $\boldsymbol{\lambda}_j$ varies with perturbations in the inputs.

5.4. Example of a Sensitivity Analysis of the MD River Basin Model

Four sets of inputs and parameters were varied, each at two levels, in a sensitivity analysis of the MD river basin. The inputs and parameters are listed in Table 5.1, together with their values at both levels. In all, sixteen trials were involved.

Table 5.1

Variations in Model Inputs

Input	Level 1	Level 2
Interest rate, r (%)	0.04625	0.050
Reservoir costs, RC	Costs are listed in Table 4.1	110% of costs of level 1
Hydrology, HY	Mean flows from historical hydrological record	90% of mean flows
Demand profiles, DP	Irrigation demand doubles in 10 yr and then remains constant	Irrigation demand doubles in 7 yr and then remains constant

Failure to meet irrigation demands was introduced by a linear loss function that penalized large violations of the demand constraints. The penalty function associated with arc (5, 2), $PF_{5,2}$, took the following form:

$$PF_{5,2} = 20{,}000(HI_{5,2} - F_{5,2})$$

where HI is the capacity of the arc, and F is the flow in the arc. The penalty function reflects the increasing damage to unirrigated and fallow lands. Similar penalty functions were associated with the other irrigation arcs and may be incorporated in the return function.

The results of the sixteen sets of trials are summarized in Tables 5.2 and 5.3.

In eleven of the cases, the dam construction schedule was insensitive to perturbations of the inputs and parameters. In each of the eleven cases, two irrigation reservoirs were constructed first and then a hydroelectric reservoir was built. However, when the interest rate and the demand schedules were both simultaneously at their lower levels, the scheduling sequence was changed. The significant change that took place was that the hydroelectric dam was assigned second priority rather than third because there was a satisfactory trade-off between increased energy revenues and the irrigation losses caused by the delay in constructing another irrigation dam. However, over the *short term* the same dam (number 9) was chosen to be built first. The decision maker thus could wait until additional information was available on demand profiles before deciding which dam to build next.

Table 5.2

Variation of Net Return[a] f with Different Combinations of Inputs

		Demand profile			
		$(DP)_1$		$(DP)_2$	
		Interest rate			
Hydrology	Reservoir costs	r_1	r_2	r_1	r_2
HY_1	RC_1	150.43	139.72	151.03	140.24
	RC_2	146.34	No Solution	146.65	No Solution
HY_2	RC_1	143.44	133.06	143.64	133.28
	RC_2	138.98	No Solution	139.21	No Solution

[a] Return is in millions of dollars.

Table 5.3

Variation in Dam Construction Schedules with Different Combinations of Inputs

		Demand profile			
		$(DP)_1$		$(DP)_2$	
		Interest rate			
Hydrology	Reservoir costs	r_1	r_2	r_1	r_2
HY_1	RC_1	(9, 4)[a] (6, 9) (12, 10)	(9, 4) (6, 9) (12, 10)	(9, 3) (12, 4) (6, 10)	(9, 3) (12, 5) (6, 10)
	RC_2	(9, 4) (6, 9) (12, 11)	Infeasible	(9, 4) (12, 5) (6, 11)	Infeasible
HY_2	RC_1	(9, 3) (12, 4) (6, 10)	(9, 2) (12, 4) (6, 10)	(9, 2) (12, 4) (6, 10)	(9, 2) (12, 4) (6, 10)
	RC_2	(9, 2) (12, 4) (6, 11)	Infeasible	(9, 2) (12, 4) (6, 11)	Infeasible

[a] Group of three items signifies build dam 9 in year 4, dam 6 in year 9, and dam 12 in year 10.

When both the interest rate and the reservoir costs were raised simultaneously, no solution was possible for scheduling the dam construction. The lack of a feasible solution was caused by the fact that construction of reservoirs 9 through 14 would *not* yield a positive net benefit even if the reservoirs were operated at the highest efficiency over all their useful life. For a real problem, if similar results occurred, the cost data for these reservoirs should be scrutinized more closely before making any decisions about reservoir construction.

In summary, this section points out the need for testing the sensitivity of the outputs of a model to input and parameter variations before making any final decisions, particularly in the construction of reservoirs that require large investments.

5.5. Stochastic Analysis of a Water Resources System

A major assumption made in the formulation of the mathematical model of the water resources system discussed in Chapter 4 was that all the system inputs and parameters, such as the hydrological inputs, future demand profiles, and reservoir costs, were deterministic. Although deterministic analysis leads to useful results, most of the significant water resources parameters are in truth stochastic in nature because of the uncertainty in rainfall distributions, population growth, and future economic conditions. A few brief remarks on some of the routes to stochastic water resources planning are appropriate at this point insofar as they attempt to answer the same questions that are posed in a sensitivity analysis.

If simulation is to be used in the planning for a water resources system, three successively more comprehensive strategies can be employed with a stochastic runoff, stream flow, demand, and so on: (1) simulation analyses using only historical sequences of data, (2) simulation analyses using a number of possible future sequences of data, and (3) simulation followed by deterministic optimization followed by regression.

Method 1 presumes that the sequence of stream flows, runoff, water demands, and so on that occurred in the past will occur in the future. It leads to operational policies that are based on the assumption that the worst that has occurred historically will occur once again. Method 1 was employed long before the current digital computational capability existed, and would be optimal only if the expectation of penalties for the worst event greatly outweighed all other considerations. Method 1 subsumes the use of "critical period hydrology" a technique that has been greatly favoured by Hall and co-workers (see Section 2.1).

Method 2, which can be characterized as ordinary simulation, was developed by members of the Harvard Water Program in the early 1960s. Stream flow, runoff, and similar data are fitted by probability distributions and functions of time and distance. Then these relationships together with a random-number generator can be used to generate equally likely future data sets. Operational decisions, target levels, and physical configuration are set, and a simulation study is run for a number of possible future sequences. The main difficulty with method 2 is not so much that it does not adequately represent reality but that the results generated may not be conclusive. Four levels of simulation are present: hydrology, operational policy, commitment level, and physical facilities. The decisions to be reached regarding operational policy and physical facility configuration and timing are highly dimensional. To adequately span the population of possible ways in which various future system inputs might occur, 30 or 40 sequences may be barely adequate, and much larger number of sequences, say 200, is likely to be needed for statistical reliability. Thus, as straightforward as method 2 may seem, it is usually too costly and may even be infeasible from a computational standpoint.

Method 3 differs from method 2 in that one tries to restrict the extent of the simulation somewhat by intermediate optimization for the operation policy rather than supplying an operation policy and examining the result. A general outline of method 3 is:

1. Simulate a possible future set of stochastic data.
2. Find the optimal operating policy for that set of data.
3. Repeat steps 1 and 2 several times.
4. Carry out a regression analysis on the results to estimate an operating policy that is optimal given that there are several possible future sets of data for stream flow, runoff, and so on. The approach of method 3 was also pioneered at Harvard University by Young and Fiering who developed reservoir operating rules using forward dynamic programming in step 2. They christened their technique Monte Carlo Dynamic Programming (Young, 1967).

Figure 5.3 illustrates one way to select the stochastic sequences to be used.

The same model and algorithm described in Chapters 2 and 3 may be used for determination of capital investment strategies and the degree of associated risk for systems with stochastic inputs by method 3 as follows.

For the planning problem, including stochastic parameters, the capital investment optimization algorithm is combined with Monte Carlo experimentation and duality analysis to determine the optimum river basin management strategy and the risks involved. A river basin can be chosen to have a current configuration of unregulated streams and rivers, of reservoirs, of canals, and of sites for future additional construction. From the hydrological record a series of representative synthetic flow sequences can be generated by an equation such as the Thomas–Fiering equation [Fiering, 1966] and used as inputs to the model. Given a unique set of costs, interest rates, and future profiles of demand and pollution load, an optimal solution can be found for each hydrological sequence, as well as (1) the expected return from operating the system, (2) the risk of not meeting the demands on the system, (3) the sensitivity of construction schedules to variations in the hydrological flow sequences, and (4) whether the *sizes* of the structure built were truly optimal, as ascertained through a duality analysis.

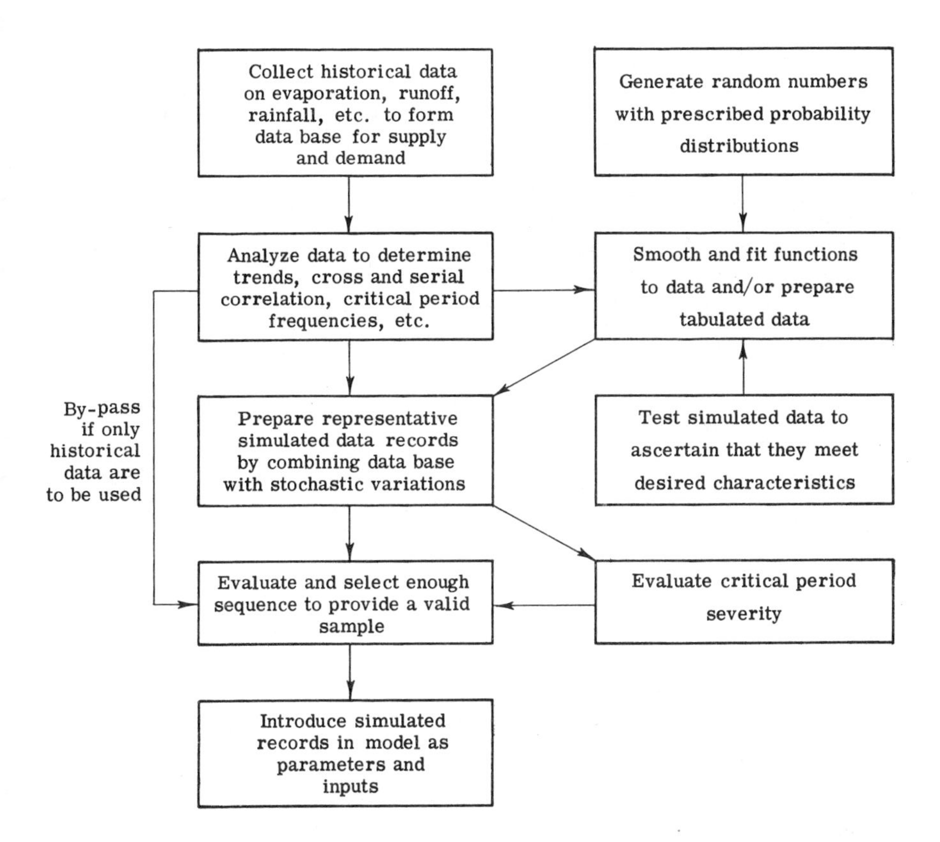

Fig. 5.3 How to prepare and introduce stochastic variables into deterministic models of water resources systems.

A completely different method of treatment of the stochastic problem is to directly use a stochastic optimization technique, such as chance-constrained programming or stochastic dynamic programming. However, discussion of these methods falls considerably outside the scope of this book.

5.6. Summary

Water resources management occurs in a dynamic environment in which changes in inputs, costs, and objectives occur because of related technological, political, and social changes. The model used in Chapter 4 is deterministic, and the sensitivity of its objective function and outputs to small variations in four inputs and parameters (each at two levels) was measured. The inputs and parameters chosen were (I) system hydrology, (II) reservoir costs, (III) future irrigation demand, and (IV) interest rate. A linear penalty function was used to punish noncompliance with irrigation demands.

In all, sixteen sets of trials were run resulting in no change in the dam construction schedule for 11 of the cases. When both the interest rate and the demand schedules were simultaneously at their lower levels the construction schedule was altered for the two later dams, but the same dam was chosen to be built first. In four cases no solution was possible for scheduling the dam construction because no positive net benefit accrued from the construction of dams 9–14.

Finally there is a discussion of stochastic analysis of water resources systems. Three strategies, all based on simulation, are discussed. It is concluded that method 3, which combines optimization and regression analysis with simulation, is the most reliable available.

Chapter 6

HOW TO INCORPORATE WATER QUALITY AND POLLUTION CONSIDERATIONS INTO THE MODEL OF THE WATER RESOURCES SYSTEM

This chapter describes how to include planning for water quality management into the previously developed scheme for the optimal expansion of a water resources system. Water quality management was omitted from the water resources planning framework described in Chapters 2 and 3 in order to simplify the explanation, but the interactive nature of a water resource system together with the fact that water in typical rivers is used and reused numerous times in its transit down the river requires a detailed examination of water quality in conjunction with the water quantity. Interestingly enough, in spite of the principle that water quality management should be an integral part of water resources planning, water quality management and water development activities are usually separated and often delegated to different agencies in the state and federal governments. As an example, in the state of Texas, water quality control problems come under the surveillance of the Texas Water Quality Board, while water development activities are handled by the Texas Water Development Board. However, along with the large increase in the demand for water from various users has come an awareness that the problem of the degredation of water quality in the future meshes directly with that of water quantity.

We will first consider the general objectives in planning for water quality and define a few new terms before proceeding to the formation of the model for water quality.

6.1. Objectives in Planning for Water Quality

In considering the objectives associated with water quality management we must distinguish, as we did in Chapter 1, between operation of a water resources system as it exists and planning for the expansion of a water resources system including water quality. In water pollution control, as in other areas of public or social investment, there may be important objectives other than strict economic efficiency. In fact in many instances, social or political factors may be dominant. However, we will be concerned here with those features that can be quantified. Furthermore, the formulation of the system model will be deterministic rather than stochastic, and, as expected, the representation of the component subsystems will not be as detailed as required for the analysis of the operation of an existing system over a short time period (relative to the planning horizon).

Without the evaluation of water quality control benefits, neither systems analysis nor comprehensive river basin planning can lead to economic justification of proposed water quality control measures. An examination of various sources that have considered the bases for the evaluation of benefits [U.S. Congress, Senate, 1962; Caulfield, 1967; White *et al.*, 1969; Kalter, *et al.*, 1969] indicates that four main objectives exist: (1) national economic efficiency, i.e., how to increase the national income and product; (2) preserving and improving the national environment for man's use and enjoyment, which may be referred to as conservation; (3) regional development; (4) intangible benefits related to scientific, historical, and cultural values. How to quantify these objectives is still a matter for research.

6.2. Concepts Involved in Planning for Water Quality

We need to define a few terms in this section that will be employed in the formulation of the quality aspect of the model of the water resources system if it is to be compatible with the model in Chapter 2. The major variables of interest in the characterization of water quality are (1) biochemical oxygen demand, (2) chemical oxygen demand, (3) nutritient materials other than carbon, (4) temperature, and (5) dissolved oxygen. We shall ignore temperature effects in what follows.

The biochemical oxygen demand (BOD) is the total oxygen requirement for the oxidation of biodegradable organic material contained in a waste stream, and is a rough measure of the concentration of the biodegradable waste. Since the constituents corresponding to the BOD consume dissolved oxygen (DO) in natural streams, the DO is also one of the key parameters indicating the quality of water. Consequently, the BOD–DO relationship will be the principal indicator governing the management and control of stream quality. Chemical oxygen demand (COD) indicates the oxygen required to oxidize all the material represented by the BOD plus certain other oxidizable carbonaceous material that cannot be oxidized by biological means. COD is much easier to measure than BOD, taking only a few minutes for a single measurement, whereas BOD measurements take considerably more time.

If all of the five factors listed above are in balance in the receiving water, adequate natural purification of waste inputs can take place to maintain an environment that supports aquatic life and is pleasing and not harmful to man. However, if the waste loading exceeds the natural purification capacity of the stream, offensive and anerobic conditions can result in associated fish kills, sludge banks, unsightly conditions, and unpleasant odors. Thus a systems analysis must formulate a model that represents the sources and sinks for dissolved oxygen so that a BOD–DO balance can be maintained as wastes are introduced into streams. The complexity of the model depends largely on how many sources and sinks for dissolved oxygen are included in it, and the time period of interest.

Numerous methods of controlling water quality exist or have been proposed. Typical techniques include direct waste treatment, water treatment, in-stream aeration, and flow augmentation. Incorporation of each method adds a further dimension of complexity to an already complex planning problem. In waste water treatment plants a certain fraction of the waste, say 30–50%, is removed by primary treatment; up to 90–95% can be removed by secondary treatment, and an even higher percentage removed by what is termed tertiary treatment. After treatment the purified waste water is discharged into a receiving stream. The present design procedure for waste treatment plants includes consideration of the characteristics of the waste water inflows, quality requirements for the treated effluent in terms of the receiving stream quality specifications, specific requirements imposed by regulatory agencies, and the reliability of the waste treatment operation. Usually the cost of waste water treatment for each specific waste treatment facility can be expressed as a function of the quality of the raw water, the quantity of the water to be treated, the quality requirements for the treated water, and the treatment process alternatives considered for the treatment plant. Regulatory agencies specify the minimum DO requirements in the stream or the maximum allowable concentration of BOD in the treatment plant effluent.

Water treatment plants will not be specifically included in the model developed in Section 6.3 because the waste water treatment model can be used for water treatment plants as well. The quality requirement for the treated water is normally specified by the usage to which the water is to be put. Municipal water has different specifications for the dissolved and suspended constituents than does industrial water. However, a cost function can be developed that depends only on the quality of the raw water at the intake of a plant if the degree of treatment is prescribed for each water treatment plant.

Regional water treatment implies the use of treatment plants to handle the waste outfalls of numerous point sources of pollution, away from their discharge locations. Obviously, economics of scale justify the construction of regional treatment plants. Historical, legal, institutional, and economic aspects of regional water quality management have been discussed by Kneese (1964, 1967). Regional water quality treatment subsumes the use of bypass piping which transfers the waste outflow of a point source to a less polluted reach of the river basin instead of dumping that waste in the most adjacent reach which is heavily polluted. In effect bypass piping is a special case of regional treatment where the treatment plants do not alter the waste content. Thus piping flows are allowed from each polluter to each river section, from each polluter to each treatment plant, and from each treatment plant to each section.

However the use of regional water treatment has not been accepted as yet in the United States mainly due to the difficulty of answering the question: who pays how much for the construction and upkeep of regional water quality plants. This question has been satisfactorily answered in Germany, for example, where the Genossenschaften (River Basin Water Quality Management Boards) have built and managed regional water quality plants with outstanding success for over 70 years. The options of regional water quality management and bypass piping have not been explicitly entered in the model, but it can be altered to include these possibilities if required.

Another means of water quality control is the use of stream flow regulation or augmentation, that is, the release of a quantity of higher quality water from an upstream reservoir to maintain or improve the quality of water in a given stream by dilution. Costs associated with the quantity and quality of the water released can be identified. The final means of maintaining water quality that was mentioned, namely in-stream aeration, involves the use of mechanical means to increase the diffusion of oxygen into the stream; this method is not widely used, and will not be included in the model.

Nearly all the water quality management models that have appeared in the literature have been concerned with the short-term aspects of water quality, that is, daily or weekly operations. Optimization has been mainly concerned with the specification of the location of water and waste water treatment plants on a river such as shown in Fig. 6.1 and their respective

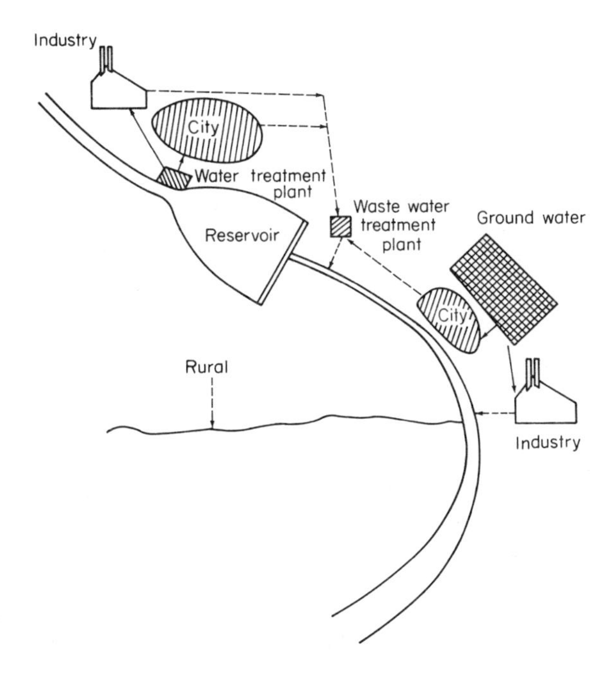

Fig. 6.1 Water and waste water sources and return in an existing water resources system. Solid lines represent water supply, broken lines represent waste water return.

levels of treatment to meet existing pollution loads at minimum cost. One-dimensional steady-state and unsteady-state models have been used to predict the BOD, DO, and temperature in river reaches, using a deterministic hydrology. Most of the models have used linear time-discrete constraints with linear objective functions. Both linear programming [Deininger, 1965a,b; Day *et al.*, 1965; Kerri, 1965, 1967; Loucks *et al.*, 1967; Revelle *et al.*, 1967, 1968; Sobel, 1965; Thomann and Sobel, 1964; Thomann and Marks, 1966] and dynamic programming [Liebman and Lynn, 1966; Dysart and Hines, 1969] have been used to solve the linear models. Graves *et al.* (1969) have developed a water quality management model using bypass piping as an added option within the framework of linear programming. The model was implemented using semirealistic data for the Delaware Estuary. When only plants with specified levels of BOD removal can be built, such as low (20%) primary removal, high (35%) primary removal, and so on, the problem becomes an integer programming one because all variables must take on integer values only. Four examples illustrating the solution of such integer problems are Deininger [1965b], Liebman [1969],

Liebman and Marks [1968], and Fitch *et al.* [1970]. Graves *et al.* (1972) have constructed a highly complex nonlinear water quality management model that allowed as decision options: (I) regional water treatment (subsuming the use of bypass piping), (II) on-site treatment plants, and (III) low-flow augmentation. Results of an application of the model to the West Fork White River in Indiana were presented.

The lack of adequate water quality and cost data hampers the development of suitable models for water quality planning. Adequate water quality data, particularly relating to the organic constituents, are rather scanty except for general references prepared by state agencies [National Engineering Co.; Texas Water Development Board, 1968; U.S. Study Commission—Texas, 1961]. Costs for waste treatment operations are also available only on a limited scale [U.S. Dept. of Health, Education, and Welfare, Public Health Service, 1963] although consulting engineering firms have proprietary data available.

Recently three papers (Clough and Bayer, 1970; de Lucia *et al.*, 1974; Nunamaker *et al.*, 1974) have touched on the problem of seeking optimal strategies of capital investment in new treatment plants combined with the overall operation of water resources systems for a river basin undergoing economic growth with a consequent threat to the quality of the basin's water resources. The following sections (1) describe a model for water quality management in a river basin that can be meshed with the water quantity model developed in Chapter 2, and (2) demonstrate the applicability of the optimization algorithm described in Chapter 3 for the solution of the model.

6.3. Formulating the Optimal Expansion Problem to Include Water Quality

We assume that the economy of a river basin has a projected growth over the planning horizon (T_{max} years) with a consequent increase in the pollution load on the water resources from new and/or increased industrial plants and cities. Hourly and daily perturbations in BOD and DO will not be of concern here; instead we will use monthly averages of BOD and DO during each year. The effluents from plants and cities already in existence are being processed by treatment plants that are working near their maximum capacity, and it is clear that these treatment facilities will not be adequate to meet future pollution loads based on predictions of industrial and population growth.

In an arbitrary river basin a number of possible treatment plant sites are available for the treatment of the anticipated increased pollution loads. Figure 6.2 is a typical example. Each plant can be built to several scales (e.g., low primary, high primary, low secondary, high secondary, tertiary, etc.); each scale is considered a separate project. Associated with each project is a capital cost and an annual operating cost. With each failure to comply with minimum water quality standards is associated a penalty cost that reflects the damage to the ecology (e.g., fish kills, reduced spawning, etc.), but this penalty has not been included in the problem statement below because of the uncertainty in quantifying it. A capital budgetary constraint (limits of public expenditure from federal and state sources) exists.

Given the above general guidelines, the optimization problem becomes: For a planning horizon $T_{\max}$ and a set of alternative waste water treatment projects $(M^* - N^*)$, select a period, if any, when each project will be introduced so that the objective function will be optimized while (1) staying within the budget limit, (2) staying within limits of the treatment technol-

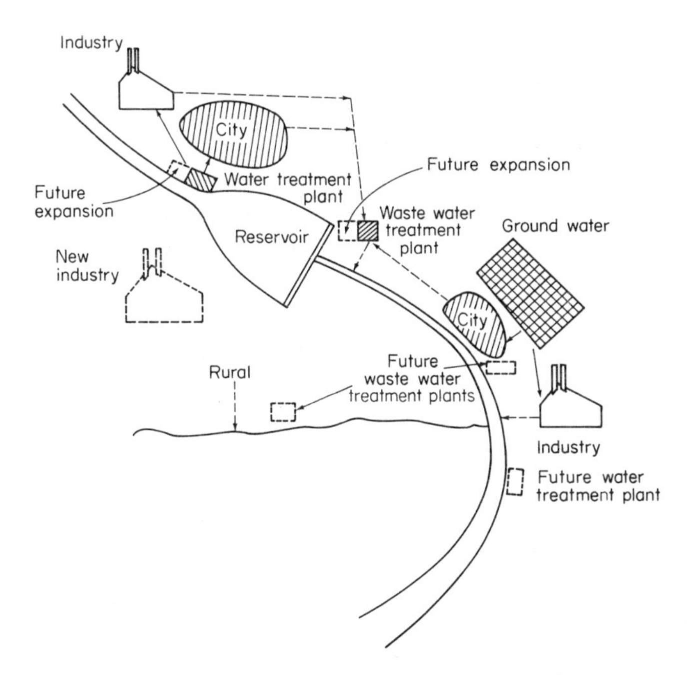

Fig. 6.2 Potential water and waste water treatment sites in the river basin, including future sources of demand and supply. Solid lines represent water supply, broken lines represent waste water return.

ogy available, (3) satisfying all physical constraints, and (4) meeting all minimum quality standards. The problem will be formulated so that the mathematical statements are directly analogous to those presented in Chapter 2 and are compatible with them.

6.3.1. Additional Terms for the Objective Function

The objective in adding waste water treatment plants is to minimize over the set of alternative projects the sum of the discounted present value of the total costs (capital and operating); hence the following terms have to be subtracted from the existing terms in the objective function (2.1):

$$f' = \underbrace{\sum_{t=1}^{T_{\max}} \alpha \sum_{j=N^*+1}^{M^*} \lambda_{jt} C_{jt}}_{\text{the } capital \text{ cost of the projects over the planning period}} + \underbrace{\sum_{t=1}^{T_{\max}} \alpha \sum_{j=1}^{M^*} \beta_{jt} \sum_{i=1}^{12} X_{ijt}}_{\text{the } operating \text{ costs for all the treatment plants}} \tag{6.1}$$

Water quality benefits, both direct and indirect, derived from higher prices for municipal and industrial water, recreational use, tourism potential, and scenic beauty, if they can be evaluated, can be added into expression (2.1) or subtracted from (6.1). There has been and no doubt will be much discussion of the magnitude and reality of these benefits, but if it is decided to include them in a planning model, they can be incorporated into the model with little difficulty if they are made functions of the volume of water flow and quality in the stream.

6.3.2. Additional Budgetary and Institutional Constraints

The budgetary constraint is dependent upon congressional and/or state water resources appropriations as in Section 2.3.1.

$$\alpha \sum_{j=N^*+1}^{M^*} \lambda_{jt} C_{jt} \leq M_t \qquad \text{for all } t \tag{6.2}$$

Constraint (6.2) exists for new projects. The additional institutional constraints are

$$\sum_{j=N^*+1}^{M^*} \lambda_{jt} \leq 1 \qquad \text{for all } t \tag{6.3}$$

i.e., at most only one new treatment plant is built in any year. Also,

$$\sum_{t=1}^{T_{\max}} \sum_{j=N^*+10}^{N^*+12} \lambda_{jt} \leq 1 \tag{6.4}$$

Constraint (6.4) is an example of an inequality that excludes those combinations of projects that are technically infeasible. It differs from constraint (6.3) because it states that only one of the three projects ($N^* + 10$, $N^* + 11$, $N^* + 12$) may ever be built. These three projects may be, for example, three different sizes of a treatment plant at a specified site.

$$\sum_{t=1}^{T_{\max}} \lambda_{jt} \leq 1 \qquad \text{for all} \quad j = N^* + 1, \ldots, M^* \tag{6.5}$$

i.e., project j can be built in only one of the $T_{\max}$ years, if it is built at all. Also,

$$\lambda_{jt} = 0 \text{ or } 1 \qquad j = N^* + 1, \ldots, M^* \tag{6.6}$$

6.3.3. Operating Cost Functional Equation

Equation (6.7) gives the operating cost function for each treatment plant as a function of the waste water flow rate QW_{ijt} and the effluent concentration BW_{ijt} from the treatment plant:

$$X_{ijt} = f_j(QW_{ijt}, BW_{ijt}) \qquad \text{for all } j \tag{6.7}$$

6.3.4. Physical Constraints

Associated with each reach m are one or more treatment plants (existing and potential). The effluents from the treatment plants are combined together and characterized by two terms, QW_{imt} (the total effluent flow for the treatment plants) and BW_{imt} (the BOD concentration of the total effluent). Equations (6.8) and (6.9) describe the total effluent flow and pollution load for an arbitrary reach. It is assumed that four treatment plants already exist in the chosen reach and three new treatment plants ($N^* + 1$, $N^* + 2$, $N^* + 3$) may be added. The same form of the constraints applies to every subsection except that summation is over different reservoirs.

$$QW_{imt} = \sum_{j=1}^{4} QW_{ijt} + \sum_{N^*+1}^{N^*+3} QW_{ijt} \qquad \text{for all } i \text{ and } t \tag{6.8}$$

$$BW_{imt}QW_{imt} = \sum_{j=1}^{4} BW_{ijt}QW_{ijt} + \sum_{N^*+1}^{N^*+3} BW_{ijt}QW_{ijt} \qquad \text{for all } i \text{ and } t \tag{6.9}$$

Inventory constraints are the constraints that describe how the DO, the BOD, and oxygen deficits vary from the beginning to the end of a reach. The material balance constraints ensure that conservation of mass is obeyed at each confluence for total water flow, DO, and BOD.

The total flow in link m is the sum of the flow in the previous link $(m-1)$, $Q_{i,m-1,t}$, the tributary flow entering the link, QT_{imt}, and the waste water flow discharged into the link, QW_{imt}:

$$Q_{imt} = Q_{i,m-1,t} + QT_{imt} + QW_{imt} \qquad \text{for all } i, m, \text{ and } t \qquad (6.10)$$

Equation (6.10) corresponds exactly to Eq. (2.13a). Assuming complete mixing, the BOD concentration at the beginning of each reach BB_{imt} is equal to the sum of the BOD concentration at the end of the previous reach, $BE_{i,m-1,t}$, that in the tributary, BT_{imt}, and that from the waste water plant(s), BW_{imt}, times their respective flows:

$$BB_{imt} = (1/Q_{imt})[BE_{i,m-1,t}Q_{i,m-1,t} + BT_{imt}QT_{imt} + BW_{imt}QW_{imt}] \qquad \text{for all } i, m, \text{ and } t \qquad (6.11)$$

In a similar fashion, the total DO at the beginning of link m, CB_{imt}, may be calculated:

$$CB_{imt} = (1/Q_{imt})[CE_{i,m-1,t}Q_{i,m-1,t} + CT_{imt}QT_{imt} + CW_{imt}QW_{imt}] \qquad \text{for all } i, m, \text{ and } t \qquad (6.12)$$

The DO deficit at the beginning of each link, DB_{imt}, is the difference between the saturation concentration, CS_{imt}, a constant, and the initial DO concentration:

$$DB_{imt} = CS_{imt} - CB_{imt} \qquad \text{for all } i, m, \text{ and } t \qquad (6.13)$$

The BOD concentration at the end of link m, BE_{imt}, is a linear function of BB_{imt}:

$$BE_{imt} = \nu_m BB_{imt} + \mu_m \qquad \text{for all } i, m, \text{ and } t \qquad (6.14)$$

Similarly the DO deficit at the end of link m, DE_{imt}, is a linear function of DB_{imt} and BB_{imt}:

$$DE_{imt} = \theta_m DB_{imt} + \gamma_m BB_{imt} + \rho_m \qquad \text{for all, } i, m, \text{ and } t \qquad (6.15)$$

The DO concentration at the end of each link, CE_{imt}, is the difference between the saturation concentration CS_{imt} and the final deficit DE_{imt}:

$$CE_{imt} = CS_{imt} - DE_{imt} \qquad \text{for all } i, m, \text{ and } t \qquad (6.16)$$

6.3.5. *Other Constraints*

(a) FUTURE POLLUTION LOADS

$$QW_{ijt} = \tilde{K}_{ijt} \qquad \text{for all } i, j, \text{ and } t \tag{6.17}$$

$\tilde{K}_{ijt}$ is the projected waste water flow from treatment plant j in month i of year t.

(b) WATER QUALITY CONSTRAINT

$$BB_{imt} \leq \sigma_m + \phi_m DB_{imt} \qquad \text{for all } i, m, \text{ and } t \tag{6.18}$$

where BB_{imt} is the maximum allowable BOD concentration (at the beginming of link m) that will not violate the maximum DO deficit for the link.

(c) TECHNOLOGY CONSTRAINTS

The upper technology constraint on BOD concentration is the maximum level of treatment attainable with present technology for secondary treatment (usually about 90% removal of BOD). As a matter of judgment the lower constraint on BOD concentration is the minimum treatment level required of each pollution source, say 35% removal of BOD.

$$BW_j^{\max} \leq BW_{ijt} \leq BW_j^{\min} \qquad \text{for all } i, j, \text{ and } t \tag{6.19}$$

6.4. Optimization of the Water Resources Model Including Water Quality

Although the water resources model of Section 6.3 and 6.4 is very similar to the problem formulated in Chapter 2, the optimization method of Sections 3.2 and 3.3 cannot be used directly to solve the combined quality and quantity problem for the following reasons.

1. In the combined problem, more than one construction decision may be made in a single year, i.e., a new dam *and* a new treatment plant could be constructed in the same year. To overcome this difficulty it would be possible to further subdivide a year into smaller periods, and it would be necessary to add the left-hand side of inequality (6.3) to that of (2.3) so that their sum is less than 1. However, a more important barrier is the following.

2. Nonlinearities appear in the OP problem. Since the reservoir releases (Q_{imt}) and the water quality (BE_{imt}) are both *independent* variables, every time that these variables are paired, nonlinear equations are formed. For example, in Eq. (6.11) the product $(BE_{imt})(Q_{imt})$ is nonlinear, as are the other terms. Similar pairings occur between the variables in Eq. (6.12).

If nonlinear functions form part of the constraints or the objective function in the OP problem, the out-of-kilter algorithm can no longer be used. Instead, the following alternative algorithm is proposed for solving the OP problem.

First, the problem sketched in Fig. 3.9 of Chapter 3 can be put into the forward dynamic programming format (Young, 1967) as in Fig. 6.3, where

$$\mathbf{S}_i = [S_{i1t}, S_{i2t}, \ldots, S_{i12t}] \qquad \mathbf{Q}_i = [Q_{i1t}, Q_{i2t}, \ldots, Q_{M_1 it}]$$

Each stage corresponds to a time period, each state vector corresponds to a set of feasible carry-over storages, and each decision vector is a set of feasible flows satisfying the carry-over storage constraints. The initial carry-over storages $\mathbf{S}_1$ are fixed.

In stage 1 (the decisionless stage) a set of carry-over storages $\mathbf{S}_2$ is assumed. The *best* set of feasible flows $\mathbf{Q}_1^*$ to satisfy $\mathbf{S}_1$ and $\mathbf{S}_2$ would be calculated by a fast nonlinear programming (NLP) code, a code especially suited to solving problems with *equality* constraints. The return r_1^* corresponding to the best set of feasible flows is also recorded. This procedure is repeated for all possible values of $\mathbf{S}_2$, and the corresponding $\mathbf{Q}_1^*$ and r_1^* values are recorded. If there are L possible values of $\mathbf{S}_2$, the NLP code is used L times.

In stage 2 a set of carry-over storages from stage 1 called $\mathbf{S}_2$ is assumed. For every possible value of $\mathbf{S}_3$, $\mathbf{Q}_2^*$ and $r_1^* + r_2^*$ are calculated by the NLP code. Another value of $\mathbf{S}_2$ is assumed and again $\mathbf{Q}_2^*$ and $r_1^* + r_2^*$ are calculated for L values of $\mathbf{S}_3$. Altogether in this and each subsequent stage L^2 calculations are needed.

For U stages the NLP code would have to be used $(U - 1)L^2 + L$ times.

In summary, the algorithm proposed for solving the OP problem is a

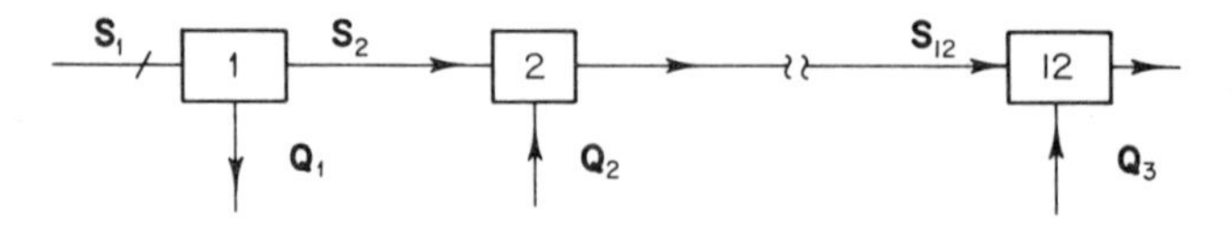

Fig. 6.3 Dynamic programming representation of a twelve-period problem.

forward dynamic programming algorithm, with an NLP code used as a subroutine. The capital budgeting problem would still be solved by the branch and bound algorithm. Clearly, the repeated use of an NLP code can be exceedingly time consuming because even the best current NLP codes require times of the order of 1–10 sec for a single execution, whereas the OKA requires times of the order of 0.01–0.1 sec for a single execution.

A more viable optimization method for the combined large-scale water management problem that includes conjunctive management of water quantity *and* water quality is a hierarchical approach similar to that described by Lasdon (1970) and Haimes (1972). The hierarchical approach requires that the management problem set up in Sections 6.3 and 6.4 be decomposed into two subproblems:

1. The water quantity problem, which is simply the problem described in Chapter 2.
2. The water quality problem, which can be stated as follows:

Maximize the comprehensive objective function [Eq. (2.1) less Eq. (6.1)] *subject to* expressions (6.2)–(6.19).

The flows used in subproblem 2 are the *outputs* of subproblem 1, while the revenue from subproblem 2 is fed back into subproblem 1 as illustrated in Fig. 6.4.

In solving subproblem 1 appropriate operation of the reservoirs (releases) assists in maintaining waste quality in the streams. Thus constraints (lower bounds) would need to be placed on the arcs in which quality must be maintained; a wide variety of combinations of water quality constraints exist. A six-step optimization strategy can be outlined:

1. Pick a set of water quality constraints.
2. Solve subproblem 1.
3. Introduce the net return from subproblem 1 and the flows generated from subproblem 1 into subproblem 2, and solve subproblem 2.
4. Determine whether the combinations of water quality constraints have been searched or eliminated. If so, stop. Otherwise, go to step 5.
5. Use a heuristic and/or branch and bound method to pick a new combination of water quality constraints.
6. Go to step 2.

The coupling variables and functions are shown in Fig. 6.4.

O'Laoghaire (1974) has solved a simpler version of the water quality problem. The formulated model is similar to the model presented in this chapter without the option of low-flow augmentation. It includes the *mini-*

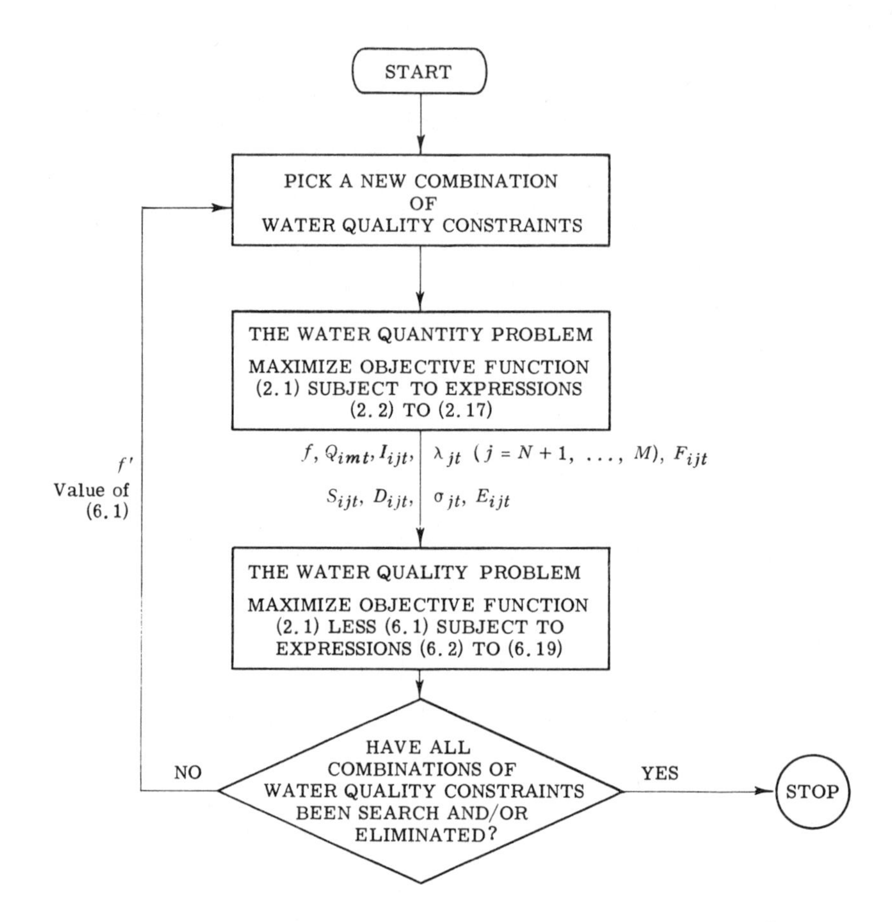

Fig. 6.4 Diagrammatic representation of the optimization method used for resolving the comprehensive water management problem of Chapter 6.

mization of the objective function Eq. (6.1) *subject to* expressions (6.2) to (6.19). The algorithm used is similar to the one explained in Chapter 3 with 2 exceptions: (a) since a minimization problem is involved the rules for picking the "best" project in any stage of the calculations will be modified to choose the least cost alternative instead of the highest return alternative and (b) a linear programming code developed for the IBM 360 Model 30 computer was used to solve the O.P. problem instead of the OKA, because equations (6.13) to (6.16) and (6.18) cannot readily be converted into network-type equations. Semi-realistic data from the Willemette River in Oregon was used to test the model.

6.5. Summary

This chapter described how to include water quality management considerations in the water quantity planning model outlined in Chapter 2. It was assumed that quality problems arose in free-flowing fresh water river reaches, and a modern modification of the Streeter–Phelps equation was used to calculate the profiles of biological oxygen demand (BOD) and dissolved oxygen (DO) in each reach of the system. Two strategies for controlling water quality were considered: low flow augmentation and direct waste treatment at the polluting source.

The formulated optimization problem became: for a planning horizon and a set of alternative waste water treatment projects select a period for introducing each project so that the objective function is minimized while meeting all the constraints (budgetary, technological, physical and water quality standards). The proposed solution approach required that the comprehensive water management problem be decomposed into two subproblems: (1) the water quantity problem of Chapter 2 and (2) the water quality problem described in Section 6.3. The flows used in subproblem 2 are the outputs of subproblem 1 while the revenue from subproblem 2 is fed back into subproblem 1.

Appendix A

COMPUTER PROGRAM LISTING AND INSTRUCTIONS

This appendix includes all the necessary instructions for running program DAMBLD, which is the computer code for the capital investment optimization algorithm outlined in Chapter 3. For convenience the appendix has been divided into six sections that deal with (1) the structure of the program, (2) roles of different routines, (3) instructions for introducing data cards into program DAMBLD, (4) block diagrams for the important routines, (5) computer program notation, and (6) a FORTRAN IV listing of program DAMBLD. The detailed output for a sample problem appears in Appendix B.

Program DAMBLD is designed to optimize the capital investment in the expansion of an existing water resources system.

A.1. Structure of the Program

The program consists of one calling routine program DAMBLD and the subroutines BAKTRAK, READIN, EXTDAT, PRINTEX, BOUND, KNOKIN(KX), SPEDAT(KX), OLERSEN, and NETFLO. The role of each routine will be discussed in the following section.

The subroutines READIN, EXTDAT, SPEDAT(KX), and PRINTEX are concerned with input/output operations; the remaining routines include all the concepts of the optimization algorithm. Block diagrams are provided for all the routines. A detailed discussion of the out-of-kilter algorithm (which subsumes subroutines OLERSEN and NETFLO) is provided in Chapter 3, Section 3.

A.2. Roles of the Different Routines

Program DAMBLD includes all the steps needed to obtain a first feasible solution to the capital investment problem. The major steps of the program are summarized in Fig. 3.14.

Subroutine BAKTRAK covers all the operations in the backtracking sequence of the algorithm; its principal operations are summarized in Fig. 3.15.

Subroutine BOUND is used to calculate the bound associated with any project at any point in time and at any stage of a construction schedule. A description of the mathematical equations used in calculating the bounds is given in Section 3.2.4. (b).

Subroutine KNOKIN(KX) is used to modify the matrix PVBAR(I,J) according to whether projects are accepted or rejected. For example, if a project S is scheduled for construction in year T, KNOKIN(KX=S) erases all values of PVBAR(I,J) for $I = S, S + 1, S + 2, \ldots, J = T + 1, \ldots,$ TMAX, where $S, S + 1, S + 2, \ldots$ are all the projects that may be built at the same site as S. Alternatively, if a project S is provisionally accepted and later rejected, the subroutine KNOKIN(KX) restores the matrix PVBAR(I,J) to its state *before* project S was considered.

Subroutines OLERSEN and NETFLO, taken together, are a codification of the out-of-kilter algorithm (OKA), which has been described in Chapter 3, Section 3.

Subroutine READIN is used to enter data about the *existing* network configuration and includes (1) the number and location of arcs and nodes and also (2) the upper bound, the lower bound, and the cost of passing unit flow for each arc.

Subroutine EXTDAT enters data about miscellaneous items and also includes data on existing and possible new reservoirs such as (1) their "active" storage, their location in the water resources network, their annual operating cost, and also (2) the cost and the maximum annual return (for new reservoirs only).

Subroutine SPEDAT(KX) is used to modify the network configuration according to whether projects are accepted or rejected. For example, if project S is scheduled for construction, SPEDAT(KX=S) changes the network configuration to accommodate the new reservoir and the new initial conditions. On the other hand, if a project S is provisionally accepted and later rejected, the subroutine SPEDAT(KX) restores the network configuration to its status before project S was considered.

Subroutine PRINTEX prints out the results from running the OKA (subroutines OLERSEN and NETFLO). The remaining output is directed by routines DAMBLD and BAKTRAK.

A.3. Instructions for Introducing Data Cards into Program DAMBLD

Table A.1 indicates the correspondence between the notation in the computer program and that in the text of the book.

A.3.1. Read-in Data

Cards 1 to N (N *is the number of existing dams.*)

On each card punch information about reservoir K.
NRES(K), the reservoir number, FORMAT I5, right justified, columns 1–5.

Table A.1

Correspondence Between the Computer Program Notation and the Mathematical Notation[a]

Computer program notation	Mathematical notation
I	j
T	t
R	r
AVCAP(T)	$\hat{C}_t$
ARCS	M_1
BND(I, T)	B_{jt}
CUMRET(I, T)	CR_t
FLOW(M)	Q_{imt}
HI	C_m
IRET(T)	$\sum_{i=1}^{12} \sum_{j=1}^{M} X_{ijt} = \hat{X}_t$
INV(I, T)	C_{jt}
LAMDA(I, T)	λ_{jt}
LO	L_m
MAR(I)	MR_j
N	N
NZC	NZC
PF(I)	PF_i
PVBAR(I, T)	OR_{jt}
TMAX	$T_{\max}$
TOTRET(T)	TR_T

[a] Units are listed in Section A.5.

NONO(K), the node number for reservoir K, FORMAT I5, right justified, columns 6–10.

CAPAC(K), the capacity of reservoirs K (in acre-ft $\times 10^4$), FORMAT F10.0, columns 11–20.

AOPCST(K), the annual operating cost for reservoir K (in dollars), FORMAT F10.0, columns 21–30.

Cards N + *1 to* MM (MM *is the total number of possible projects.*)

On each card punch information about reservoir K.

NRES(K), FORMAT I5, right justified, columns 1–5.

NONO(K), FORMAT I5, right justified, columns 6–10.

CAPAC(K), FORMAT I10, right justified, columns 11–20.

INV(K), the total capital investment required for reservoir K (in dollars), FORMAT I10, right justified, columns 21–30.

MAR(K), the maximum annual return from reservoir K (in dollars), FORMAT I10, right justified, columns 31–40.

AOPCST(K), FORMAT F10.0, columns 41–50.

LIFE(K), the economic life of reservoir K (years), FORMAT I5, right justified, columns 51–55.

Card MM + *1*

AVCAP(1), the available capital at the beginning of year 1, FORMAT F12.2, columns 1–12.

Card MM + *2*

NODES, the number of nodes in the original system (if the nodes are not numbered consecutively, use the highest node number), FORMAT I5, right justified, columns 1–5.

ARCS, the number of arcs in the original system, FORMAT I5, right justified, columns 6–10.

Cards MM + *3 to the Last Card*

I(M), J(M), the source node and the sink node for each arc M, FORMAT (8(I6, I4)), right justified.

HI(M), the maximum number of flow units that may be passed through arc M (in acre-ft $\times 10^4$, FORMAT (8(I10)), right justified.

LO(M), the minimum number of flow units that may be passed through arc M (in acre-ft $\times 10^4$), FORMAT (8(I10)), right justified.

FLOW(M), the initial flow in arc M (in acre-ft $\times 10^4$), FORMAT (8(I10)), right justified.

For each set of (I(M), J(M)), HI(M), LO(M), and FLOW(M) data, the number of cards is equal to ARCS/8 rounded *up* to the nearest integer.

In Section A.3.3 appears a listing of the data cards used for the example problem of Chapter 4.

A.3.2. Data in the Routines

A.3.2.1. PROGRAM DAMBLD

The future availability profiles are inserted beginning with the card *after* statement 180, e.g.,

$$\text{HI}(52) = \text{HI}(52) + 2$$

See the listing of the computer code.

The argument refers to the arc number and 2 is the annual increase in availability of irrigable land (expressed in terms of its water needs in acre-ft $\times 10^4$). One such statement is needed for each availability profile. If a complex series of availability profiles is used, it would be preferable to incorporate the profiles in a subroutine and use a CALL statement at this point in program DAMBLD.

A.3.2.2. SUBROUTINE BAKTRAK

The future availability profiles are inserted *in statement 120* and the following cards (one statement for each profile), e.g.,

$$120 \quad \text{HI}(52) = \text{HI}(52) + 2*(\text{T} - 1)$$

The argument refers to the arc number and T is the year in question; 2 is the annual increase in availability of irrigable land (expressed in acre-ft $\times 10^4$). If a more complicated series of availability profiles is used, follow the instructions in the preceding paragraph.

A.3.2.3. SUBROUTINE BOUND

In the Card Following Statement 50

ISUPER = the annual return from operating the *original* system (in dollars).

Sixth Card After Statement 60

AF = the annual return from operating the original system (= ISUPER).

ISUPER (or AF) may be calculated from one pass of the computer program, using subroutines OLERSEN and NETFLO.

The number of IF statements immediately after statements 120 and 260 will be equal to the number of possible new projects (MM − N).

IRA, IRB, IRE, . . . , IRZ represents the indices of different projects that may be built at any one site. In all cases the bound is calculated for the project whose argument is equal to IRA. Thus in the statements following statement 130 there are (MM − N) groups of equivalent statements, such as

```
140  IRA=7
     IRB=8
     IRE=6
```

In this case the bound will be calculated for reservoir 7; since reservoirs 8 and 6 may also be built at this site, their corresponding values of PVBAR will be ignored in the bound calculations. This is ensured by introducing below statement 220:

```
GODBAR(IRB,J)=0
GODBAR(IRE,J)=0
        ⋮
GODBAR(IRZ,J)=0
```

A.3.2.4. SUBROUTINE EXTDAT

Between the comment statements INITIAL CONDITIONS and EXISTING RESERVOIRS, the values of the variable IGORET(1) through SUMA can be inserted with any desired format. Their significance is outlined in the list of variables in Section A.5.1.

A.3.2.5. SUBROUTINE SPEDAT(KX)

The number of IF statements immediately after the statement LOGICAL OPTIME will be equal to the number of possible new projects (MM − N). Note that there are (MM − N) statements similar to the following:

```
10  IF(.NOT.OPTIME)GO TO 20
```

Between each two .NOT.OPTIME statements occurs information on the network configuration changes according to whether it is decided to add or remove a particular dam from the network.

If it is decided to add a dam to the system, OPTIME=.TRUE. and the statements immediately after the .NOT.OPTIME statement describe the changes needed to incorporate the dam in the network. If it is decided not to add the dam to the network, OPTIME=.FALSE. , and the statements immediately after the *first numbered statement following the .NOT.*

OPTIME statement describe the changes needed to revert the system to its status before the project was considered.

For example, if we consider the addition or removal of reservoir 8, we proceed as follows:

1. Go to the appropraite IF statement after the statement LOGICAL OPTIME, e.g., IF(KX.EQ.8)GO TO 50. Thus we are interested in all statements *between* statements 50 and 70.
2. If we want to remove dam 8 from the system, OPTIME=.FALSE. , and we consider all the statements between statements 60 and 70.
3. If we want to *add* dam 8 to the system, OPTIME=.TRUE. , and we consider all the statements between statements 50 and 60.

A.3.2.6. SUBROUTINE KNOKIN(KX)

The number of IF statements immediately after the statement

```
DO 60 L=T,50
```

will be equal to the total number of possibly new projects (MM − N).

Once again the indices IRA, IRB, . . . , IRZ represent the number of different projects that may be built at any one site. The number of sets of these indices will be equal to the number of dam sites (in the example in Chapter 4 there were 3 dam sites). The statements between statements 40 and 60 will have to be altered to account for the total number of projects that may be built at any one site. For example,

```
40  IF(.NOT.OPTIME)GO TO 50
    GODBAR(IRA,L) =PERBAR(IRA,L)
    GODBAR(IRB,L) =PERBAR(IRB,L)
    GODBAR(IRE,L) =PERBAR(IRE,L)
         ⋮
    GODBAR(IRZ,L) =PERBAR(IRZ,L)
50  GODBAR(IRA,L) =0.
    GODBAR(IRB,L) =0.
    GODBAR(IRE,L) =0.
    GODBAR(IRZ,L) =0.
         ⋮
60  CONTINUE
```

A.3.3. Data Input

The data input for the example problem worked out in Chapter 4 are as follows; a computer listing is provided as well.

								Calling routine	FORMAT number	Comments
1 1	40							EXTDAT	10	Data on existing reservoirs
2 4	30									
3 16	18									
4 22	30									
5 28	40									
6 14	60	4	27		50			EXTDAT	20	Data on new reservoirs
7 14	50	38	243		50					
8 14	40	34	216		50					
9 21	12	63	36		50					
10 21	10	55	30		50					
11 21	8	455	24		50					
12 27	20	10	600		50					
13 27	15	84	48		50					
14 27	10	555	30		50					
28								EXTDAT	30	Budgetary limit for year 1
240 131								READIN	10	Number of nodes and arcs
239 1	239 2	239 4	239 10	239 14	239 16	239 21	239 22	READIN	20	The set of ordered pairs (i, j) that form the arcs of the system
239 27	239 28	239 1	239 4	239 14	239 16	239 21	239 22			
239 27	239 28	1 2	2 3	3 10	4 5	5 6	6 8			
8 10	10 11	11 12	14 15	15 12	12 18	16 18	18 19			
19 26	22 24	21 24	24 26	26 32	28 30	27 30	30 32			
5 7	7 6	3 9	8 9	9 10	11 13	13 12	15 13			
16 17	19 20	22 23	24 25	28 29	30 31	1 41	4 44			
14 54	16 56	21 61	22 62	27 67	28 68	239 41	239 42			
239 44	239 50	239 54	239 56	239 61	239 62	239 67	239 68			
41 42	42 43	43 50	44 45	54 46	46 48	48 50	50 51			
51 52	54 55	55 52	52 58	56 58	58 59	59 66	62 64			
61 64	64 66	66 72	68 70	67 70	70 72	45 47	47 46			
43 49	48 49	49 50	51 53	53 52	55 53	56 57	59 60			
62 63	64 65	68 69	70 71	41 240	44 240	54 240	56 240			
61 240	62 240	67 240	68 240	32 240	72 240	17 240	20 240			
23 240	25 240	29 240	31 240	57 240	60 240	63 240	65 240			
69 240	71 240	240 239								
								READIN	30	Unit flow costs of the arcs
	- 3			- 2		- 4				
-3	-2	-4	-3	-4	-3					
							- 4			
		- 3		-5						
40	15	30	10	70	12	11	20	READIN	30	The upper flow capacities of the arcs
15	30	20	15		9		30			
	40	80	90	70	60	40	70			
60	190	190	80	80	290	25	300			
300	70	50	80	330	60	50	95			
40	15	55	40	20	250		6			
20	40	30	15	30	35	20	15			
	18		30		40	60	20			
50	30	60	25	15	40	30	60			
80	90	70	80	60	70	60	190			
190	60	60	290	25	300	300	70			
50	80	330	60	50	95	40	15			
55	40	20	250		60					
				20	15		18			
	30		40	400	400	20	40			
30	9	50		20	40	30	15			
50	35	9								
40	15	30	10	70	12	11	20	READIN	30	The lower flow capacity of the arcs
15	30	20	15		9		30			
	40									
			8				8			
						20	15			
			8		12	60	20			
50	30	60	25	15	40	30	60			
	8				8					
				20	15					
	6		12							
40	15	30	10	70	12	11	20	READIN	30	Initial flow values for the arcs
15	30	20	15		9		30			
	40	40	55	50	30	15	30			
15	95	95	70	70	165	1	166			
126	12	11	8	134	28	15	8			
15	15	5	15	20						
20	40	30	15	30	35	20	15			
			8		12	60	20			
50	30	60	25	15	40	30	60			
60	80	70	50	35	50	40	160			
160	60	60	220	7	227	227	18			
15	33	260	32	30	62	15	15			
10	10	20								
				20	15		18			
	30		40	142	322	20	40			
30	15	30	35							
		757								

A.4. Block Diagrams for the Important Routines

In the following pages flow charts are given for all the important routines of the optimization program. Figure A.1 explains the symbols used. Enough information has been given in the book and in the previous sections of this appendix to utilize the other routines effectively.

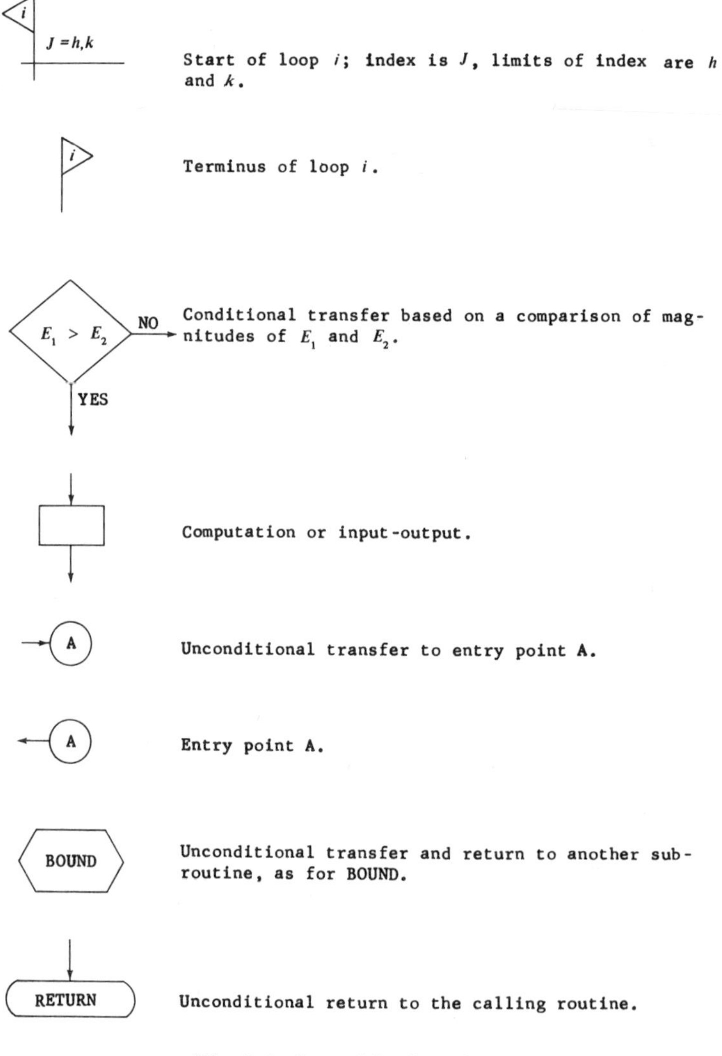

J = h,k
Start of loop i; index is J, limits of index are h and k.
Terminus of loop i.
E1 > E2
NO
YES
Conditional transfer based on a comparison of magnitudes of E1 and E2.
Computation or input-output.
A
Unconditional transfer to entry point A.
A
Entry point A.
BOUND
Unconditional transfer and return to another subroutine, as for BOUND.
RETURN
Unconditional return to the calling routine.

Fig. A.1 Legend for flow charts.

A.4.1. Program DAMBLD

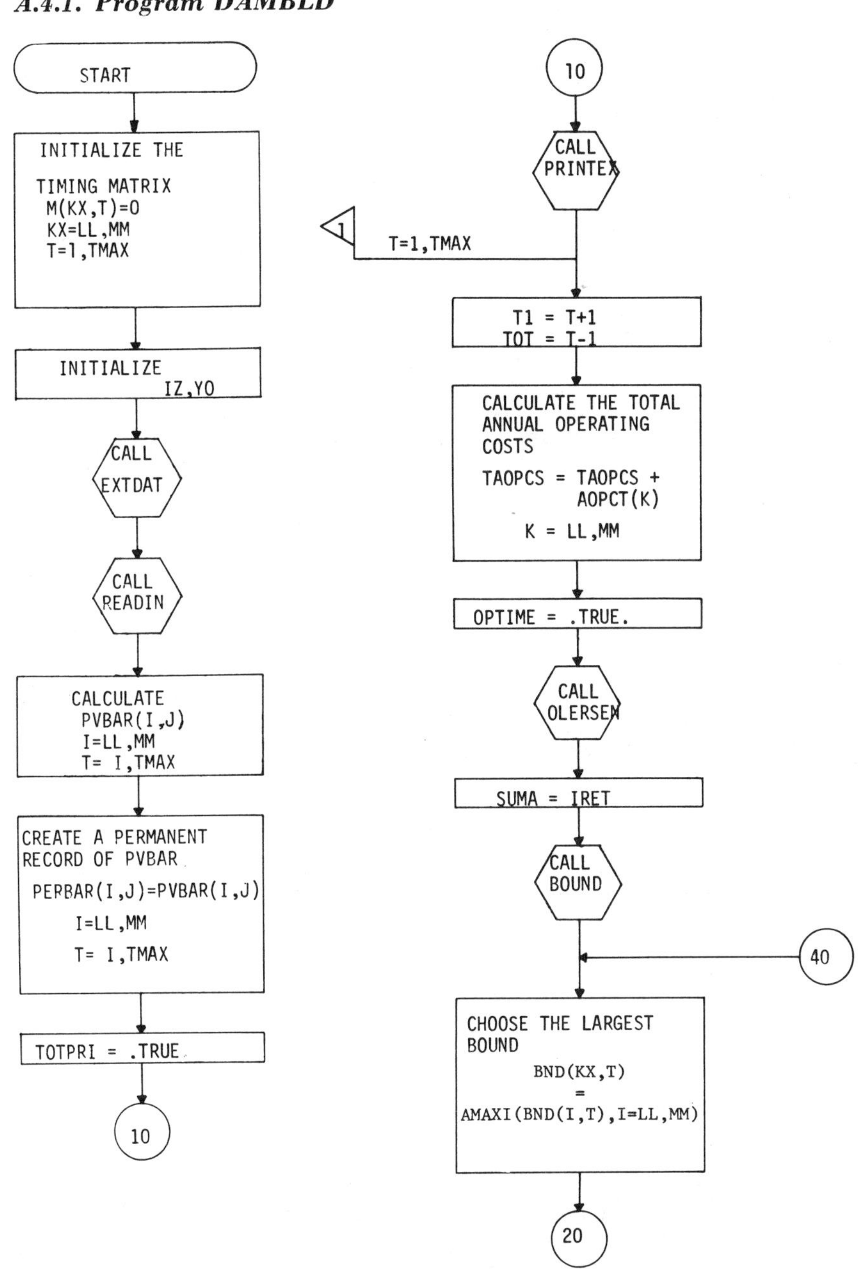

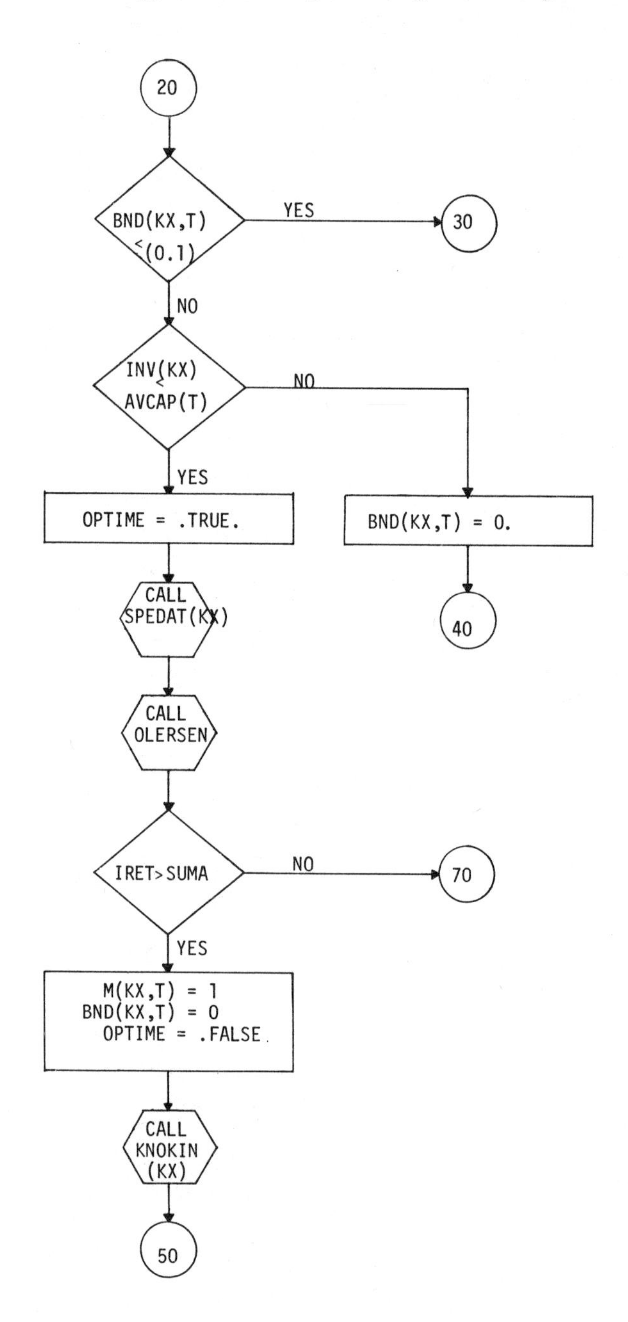
20
BND(KX,T) ≤(0.1)
YES
30
NO
INV(KX) ≤ AVCAP(T)
NO
YES
OPTIME = .TRUE.
BND(KX,T) = 0.
CALL SPEDAT(KX)
40
CALL OLERSEN
IRET>SUMA
NO
70
YES
M(KX,T) = 1
BND(KX,T) = 0
OPTIME = .FALSE.
CALL KNOKIN (KX)
50

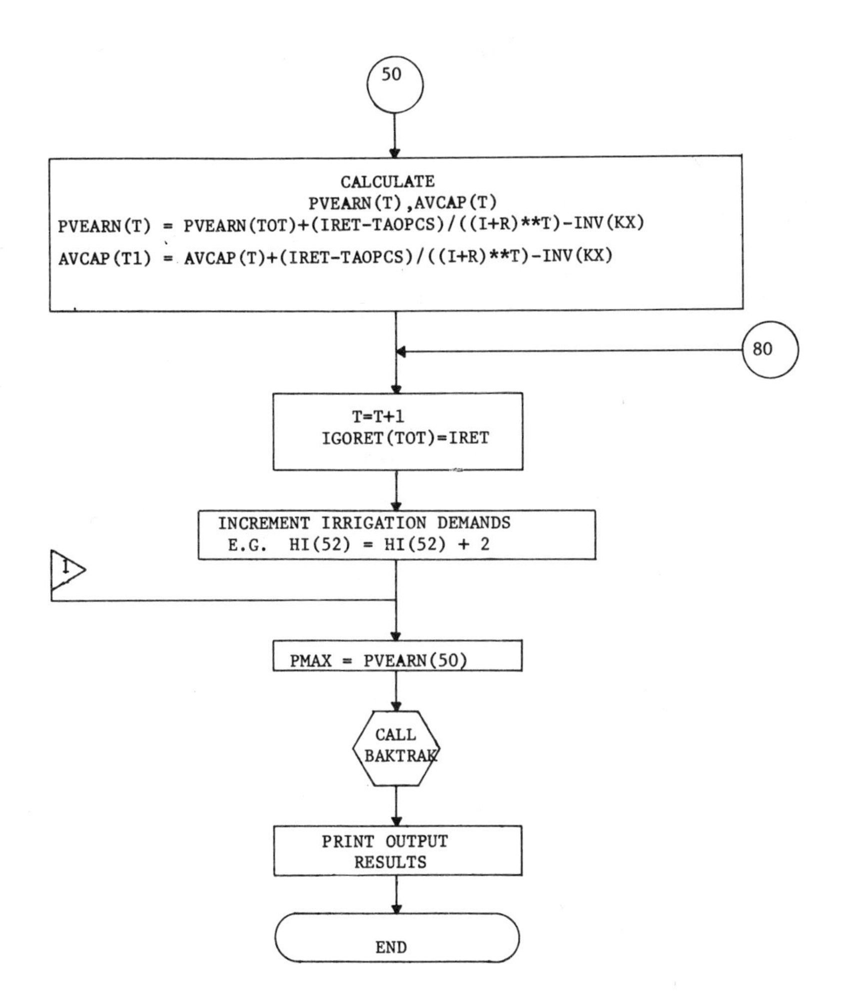
50
CALCULATE
PVEARN(T),AVCAP(T)
PVEARN(T) = PVEARN(TOT)+(IRET-TAOPCS)/((I+R)**T)-INV(KX)
AVCAP(T1) = AVCAP(T)+(IRET-TAOPCS)/((I+R)**T)-INV(KX)
80
T=T+1
IGORET(TOT)=IRET
INCREMENT IRRIGATION DEMANDS
E.G. HI(52) = HI(52) + 2
1
PMAX = PVEARN(50)
CALL
BAKTRAK
PRINT OUTPUT
RESULTS
END

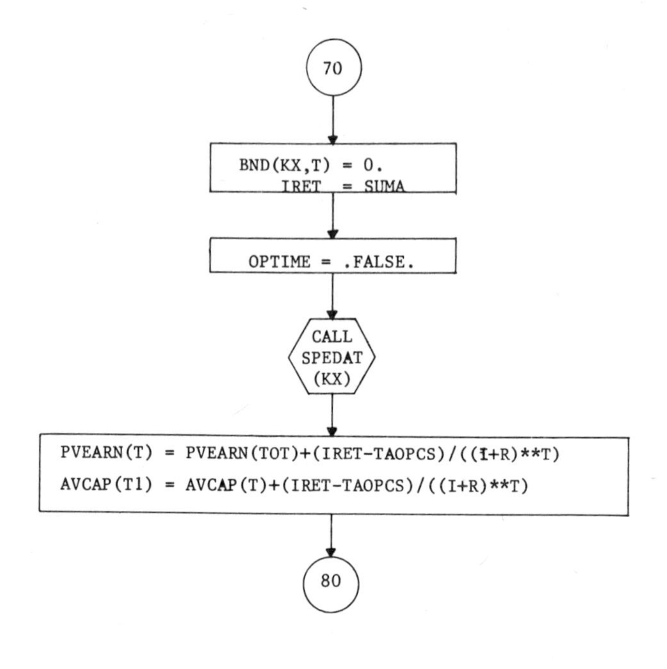
70
BND(KX,T) = 0.
IRET = SUMA
OPTIME = .FALSE.
CALL
SPEDAT
(KX)
PVEARN(T) = PVEARN(TOT)+(IRET-TAOPCS)/((I+R)**T)
AVCAP(T1) = AVCAP(T)+(IRET-TAOPCS)/((I+R)**T)
80

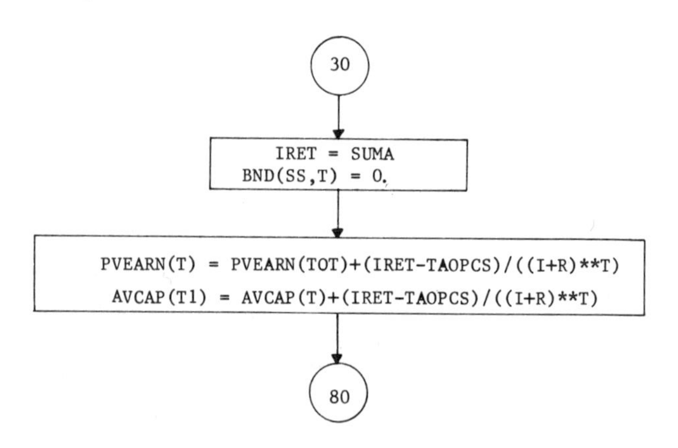
30
IRET = SUMA
BND(SS,T) = 0.
PVEARN(T) = PVEARN(TOT)+(IRET-TAOPCS)/((I+R)**T)
AVCAP(T1) = AVCAP(T)+(IRET-TAOPCS)/((I+R)**T)
80

A.4.2. Subroutine BAKTRAK

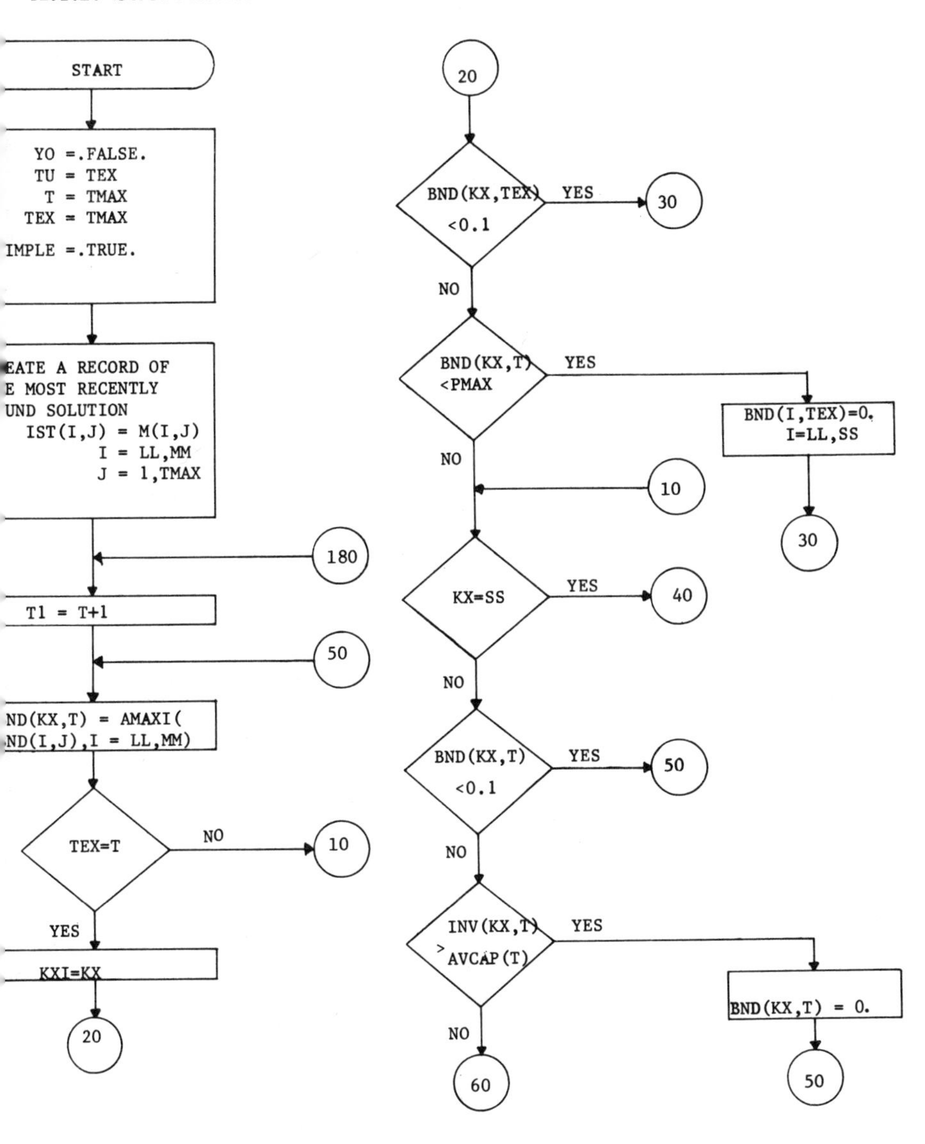

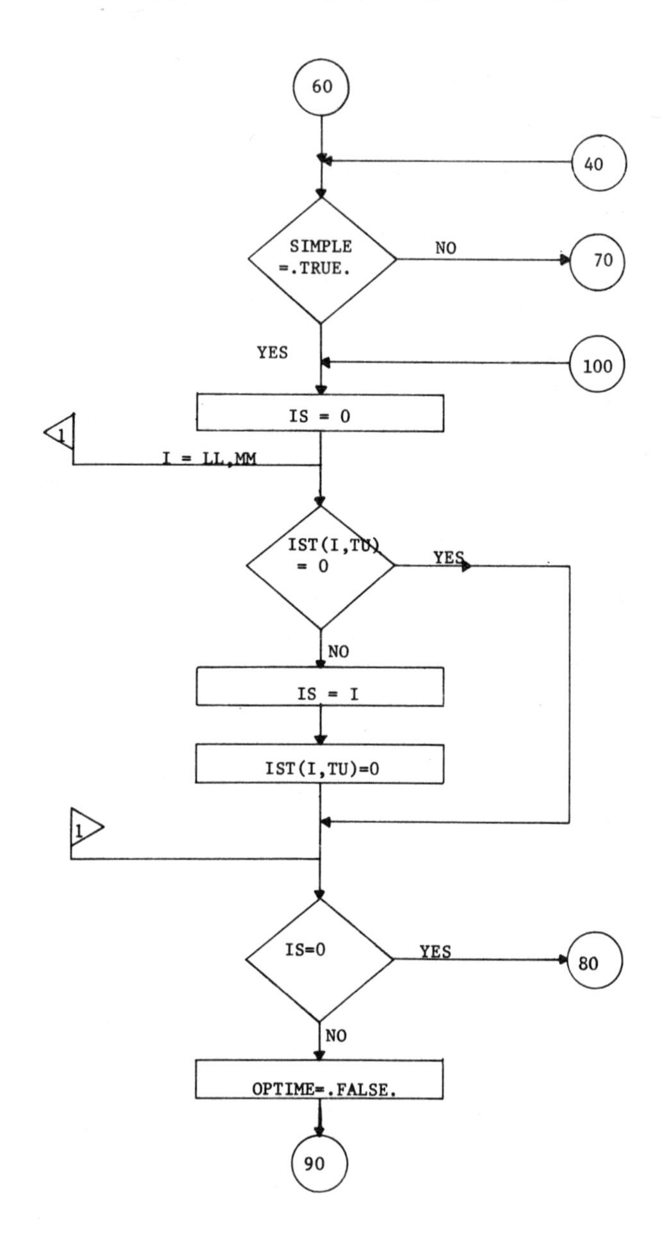
60
40
SIMPLE
=.TRUE.
NO
70
YES
100
IS = 0
1
I = LL,MM
IST(I,TU)
= 0
YES
NO
IS = I
IST(I,TU)=0
1
IS=0
YES
80
NO
OPTIME=.FALSE.
90

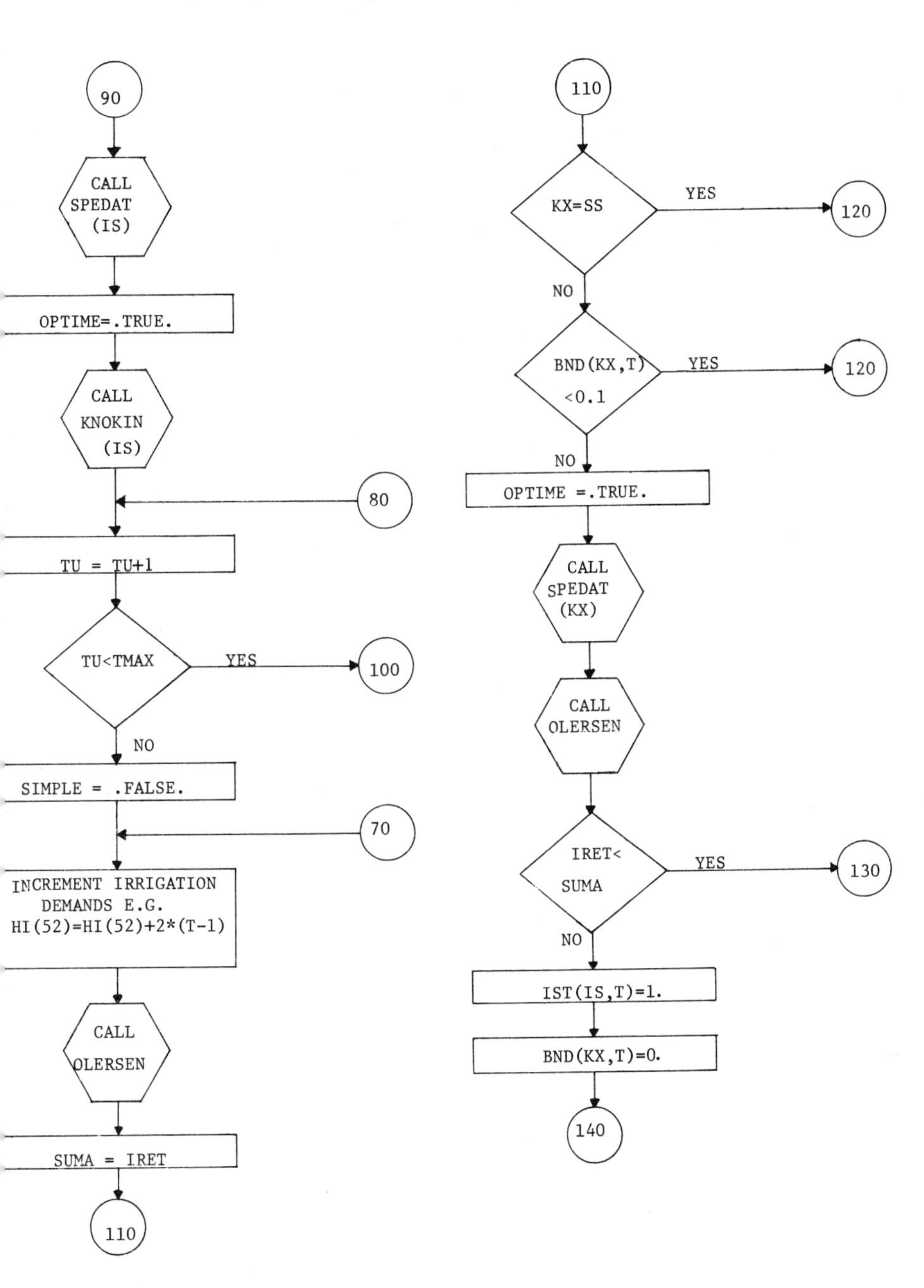
90
CALL SPEDAT (IS)
OPTIME=.TRUE.
CALL KNOKIN (IS)
80
TU = TU+1
TU<TMAX
YES
100
NO
SIMPLE = .FALSE.
70
INCREMENT IRRIGATION DEMANDS E.G. HI(52)=HI(52)+2*(T-1)
CALL OLERSEN
SUMA = IRET
110
110
KX=SS
YES
120
NO
BND(KX,T) <0.1
YES
120
NO
OPTIME =.TRUE.
CALL SPEDAT (KX)
CALL OLERSEN
IRET< SUMA
YES
130
NO
IST(IS,T)=1.
BND(KX,T)=0.
140

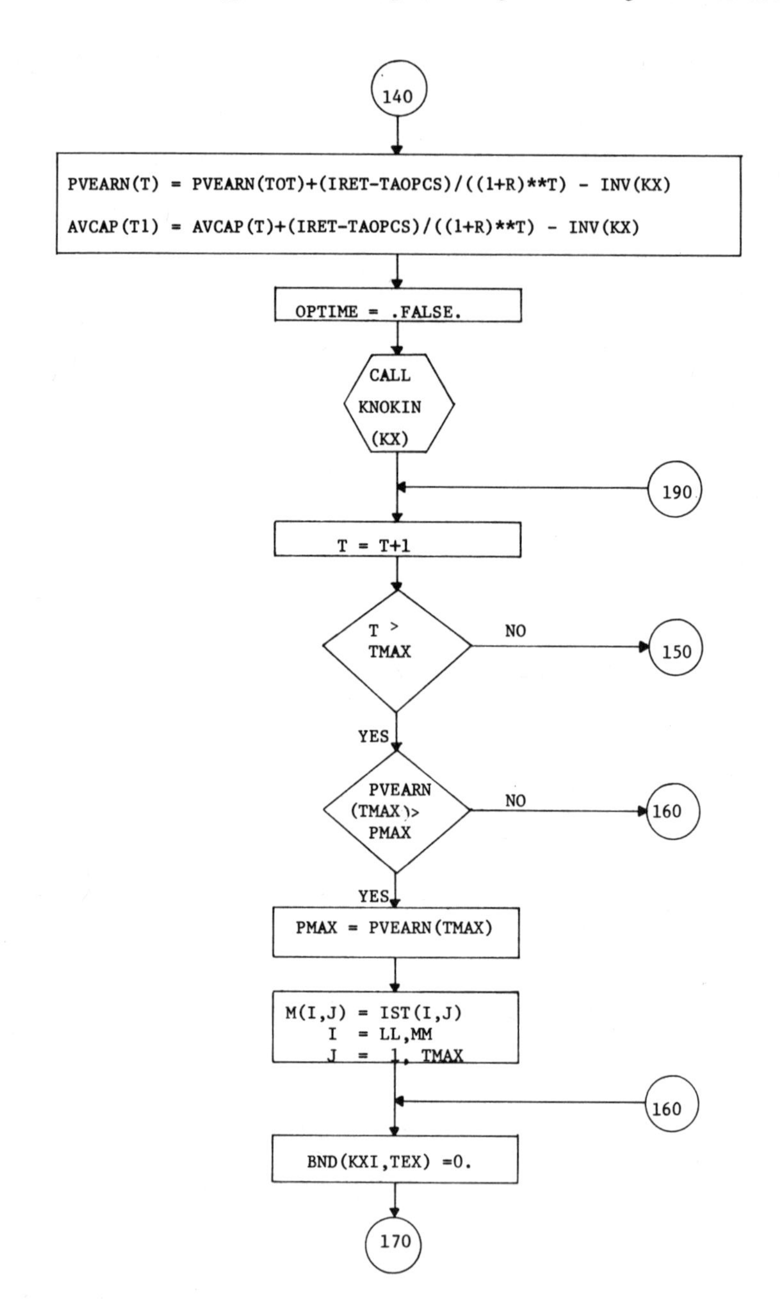

140
PVEARN(T) = PVEARN(TOT)+(IRET-TAOPCS)/((1+R)**T) - INV(KX)
AVCAP(T1) = AVCAP(T)+(IRET-TAOPCS)/((1+R)**T) - INV(KX)
OPTIME = .FALSE.
CALL
KNOKIN
(KX)
190
T = T+1
T >
TMAX
NO
150
YES
PVEARN
(TMAX)>
PMAX
NO
160
YES
PMAX = PVEARN(TMAX)
M(I,J) = IST(I,J)
I = LL,MM
J = 1, TMAX
160
BND(KXI,TEX) =0.
170

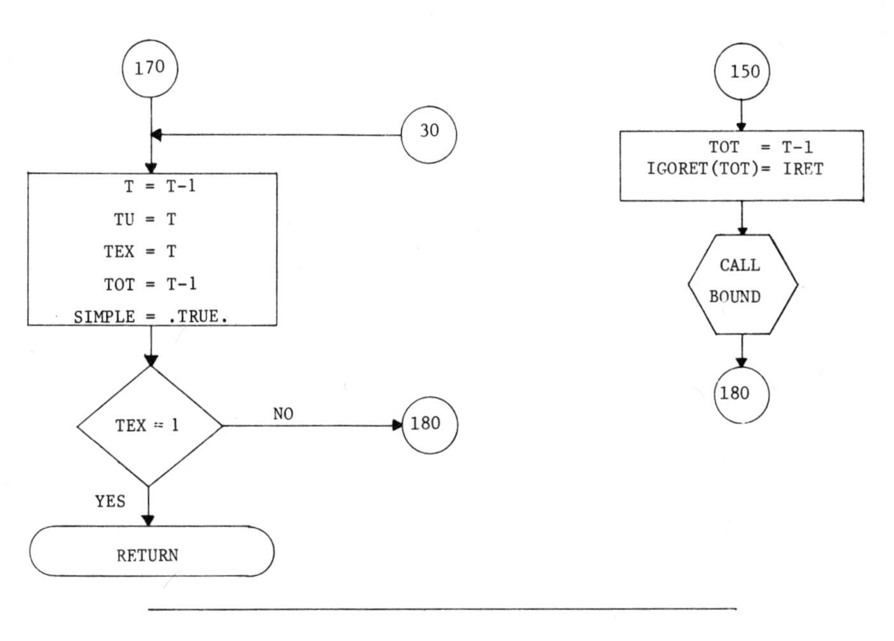
170
30
T = T-1
TU = T
TEX = T
TOT = T-1
SIMPLE = .TRUE.
TEX = 1
NO
180
YES
RETURN
150
TOT = T-1
IGORET(TOT)= IRET
CALL
BOUND
180

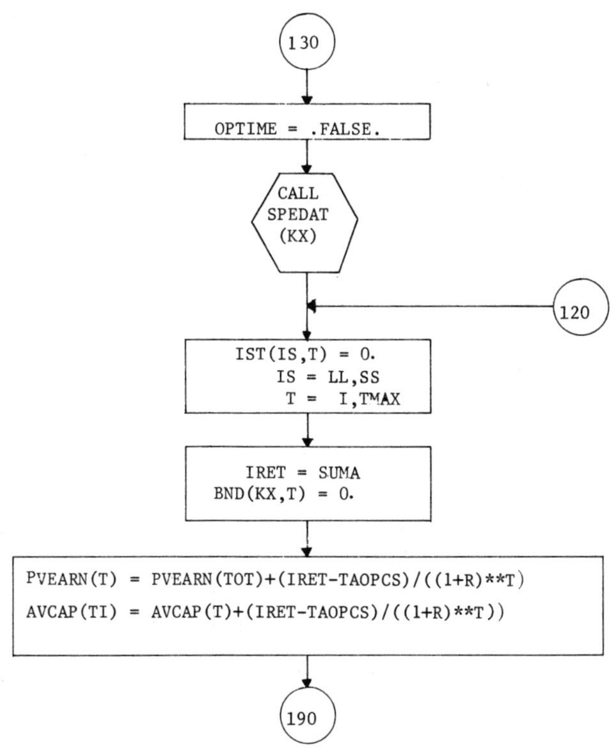
130
OPTIME = .FALSE.
CALL
SPEDAT
(KX)
120
IST(IS,T) = 0.
IS = LL,SS
T = I,TMAX
IRET = SUMA
BND(KX,T) = 0.
PVEARN(T) = PVEARN(TOT)+(IRET-TAOPCS)/((1+R)**T)
AVCAP(TI) = AVCAP(T)+(IRET-TAOPCS)/((1+R)**T))
190

A.4.3. Subroutine BOUND

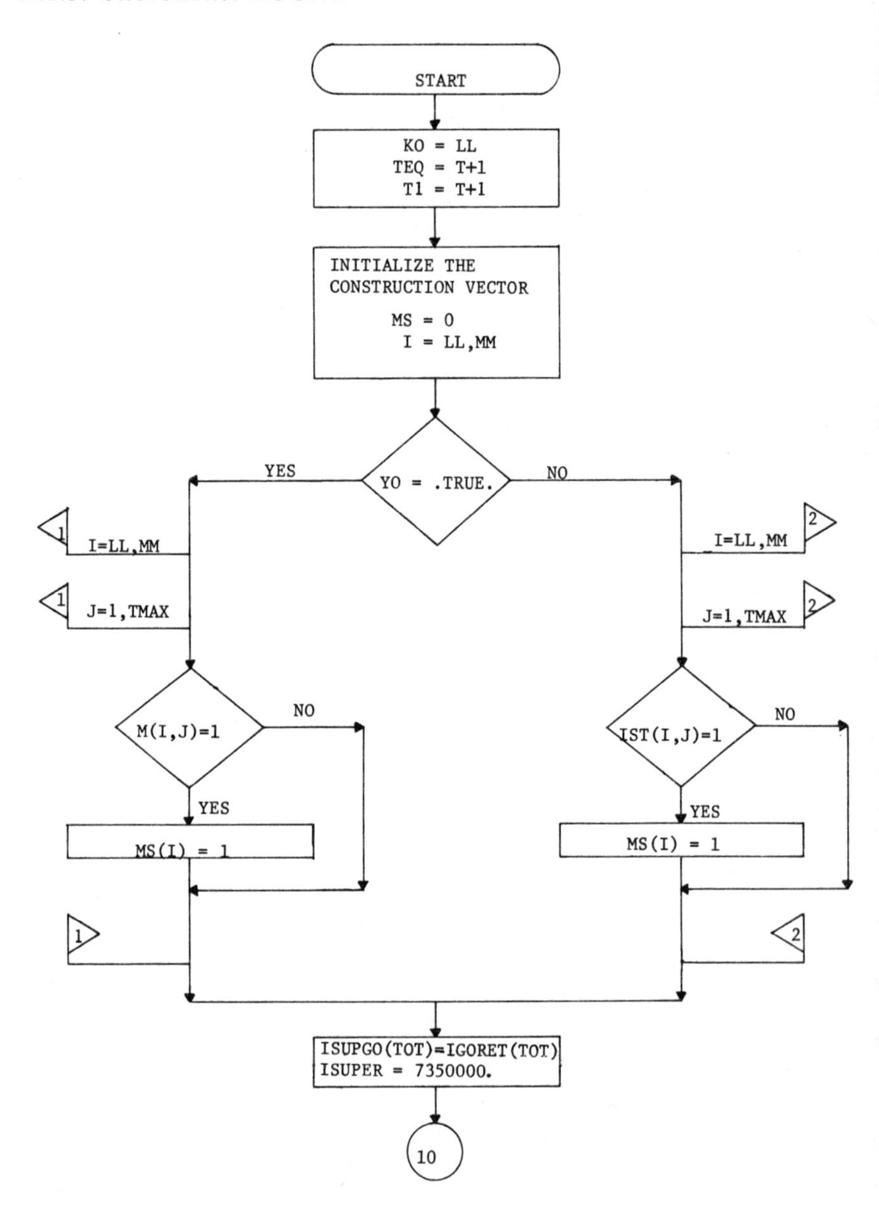

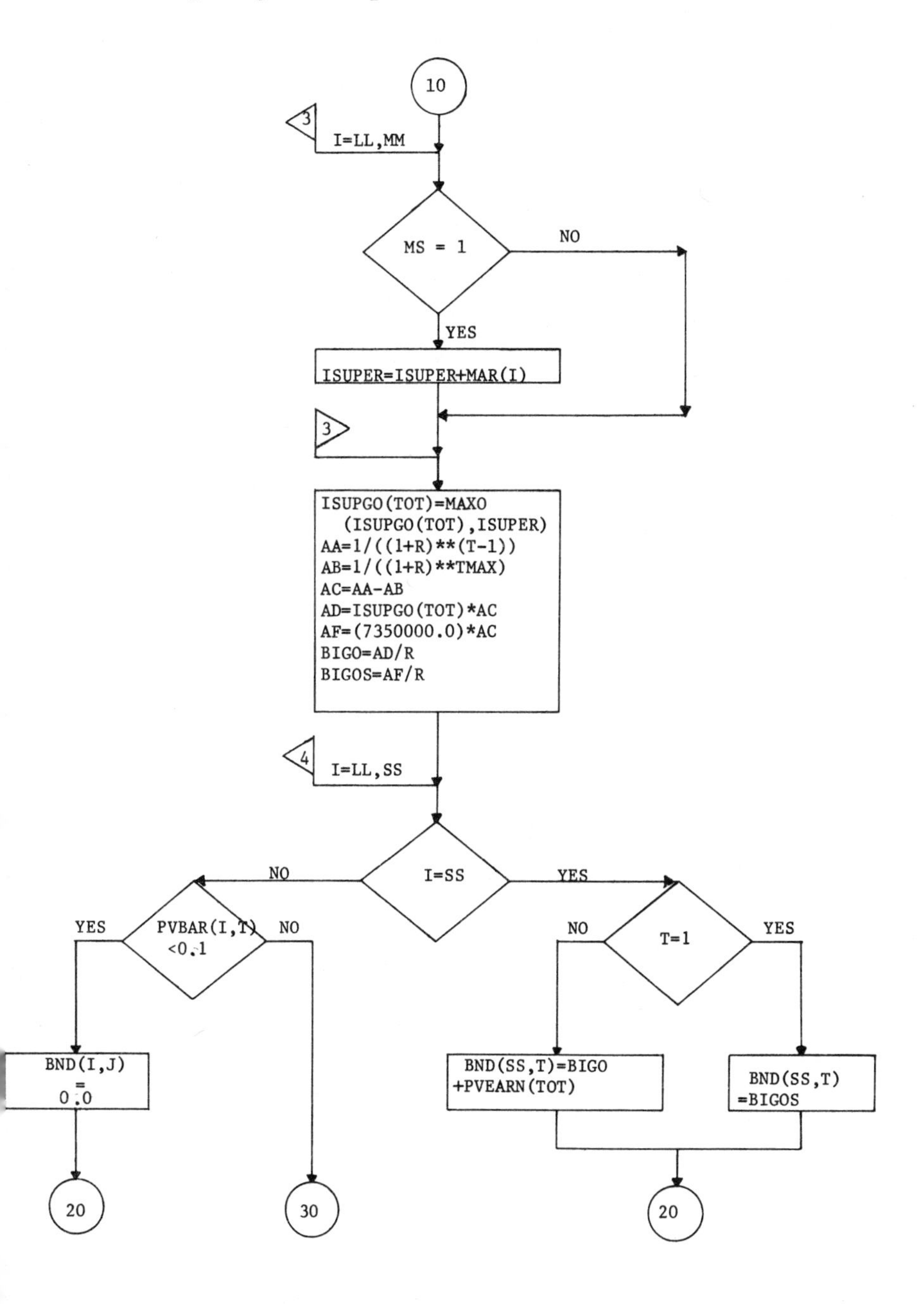
10
3
I=LL,MM
MS = 1
NO
YES
ISUPER=ISUPER+MAR(I)
3
ISUPGO(TOT)=MAXO
(ISUPGO(TOT),ISUPER)
AA=1/((1+R)**(T-1))
AB=1/((1+R)**TMAX)
AC=AA-AB
AD=ISUPGO(TOT)*AC
AF=(7350000.0)*AC
BIGO=AD/R
BIGOS=AF/R
4
I=LL,SS
I=SS
NO
YES
PVBAR(I,T)
<0.1
YES
NO
T=1
NO
YES
BND(I,J)
=
0.0
BND(SS,T)=BIGO
+PVEARN(TOT)
BND(SS,T)
=BIGOS
20
30
20

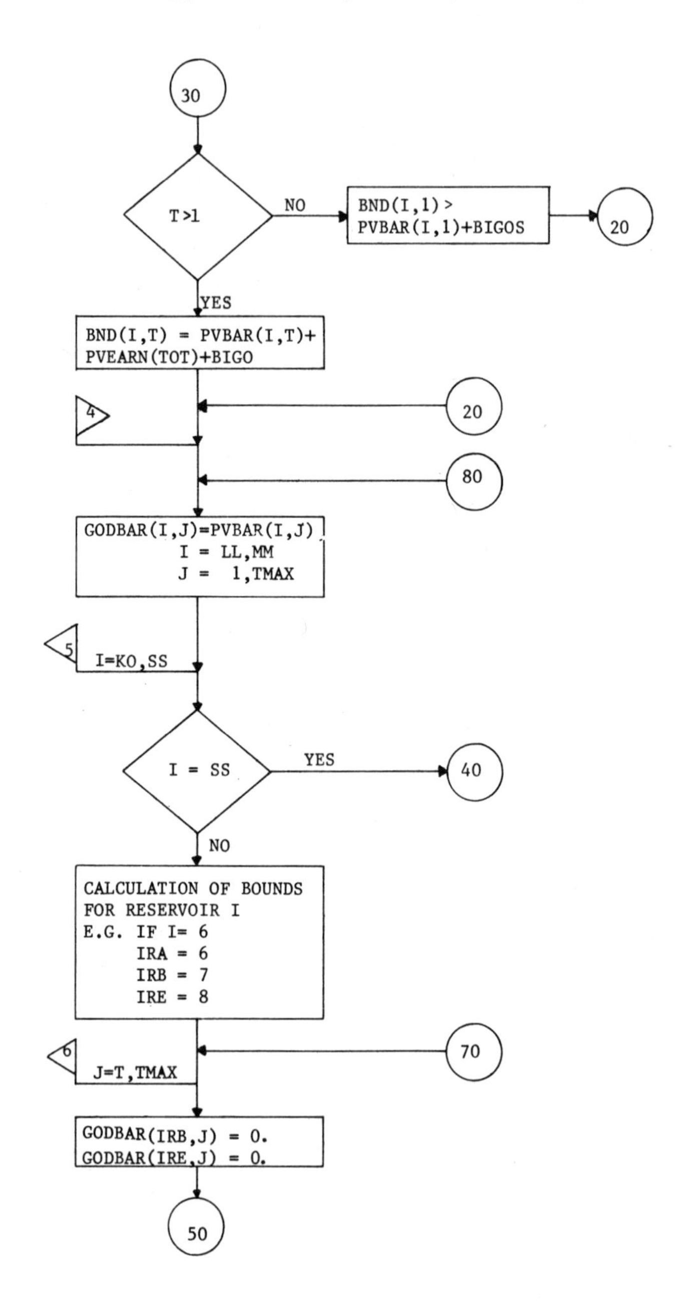
30
T>1
NO
BND(I,1) >
PVBAR(I,1)+BIGOS
20
YES
BND(I,T) = PVBAR(I,T)+
PVEARN(TOT)+BIGO
20
4
80
GODBAR(I,J)=PVBAR(I,J)
I = LL,MM
J = 1,TMAX
5
I=KO,SS
I = SS
YES
40
NO
CALCULATION OF BOUNDS
FOR RESERVOIR I
E.G. IF I= 6
IRA = 6
IRB = 7
IRE = 8
70
6
J=T,TMAX
GODBAR(IRB,J) = 0.
GODBAR(IRE,J) = 0.
50

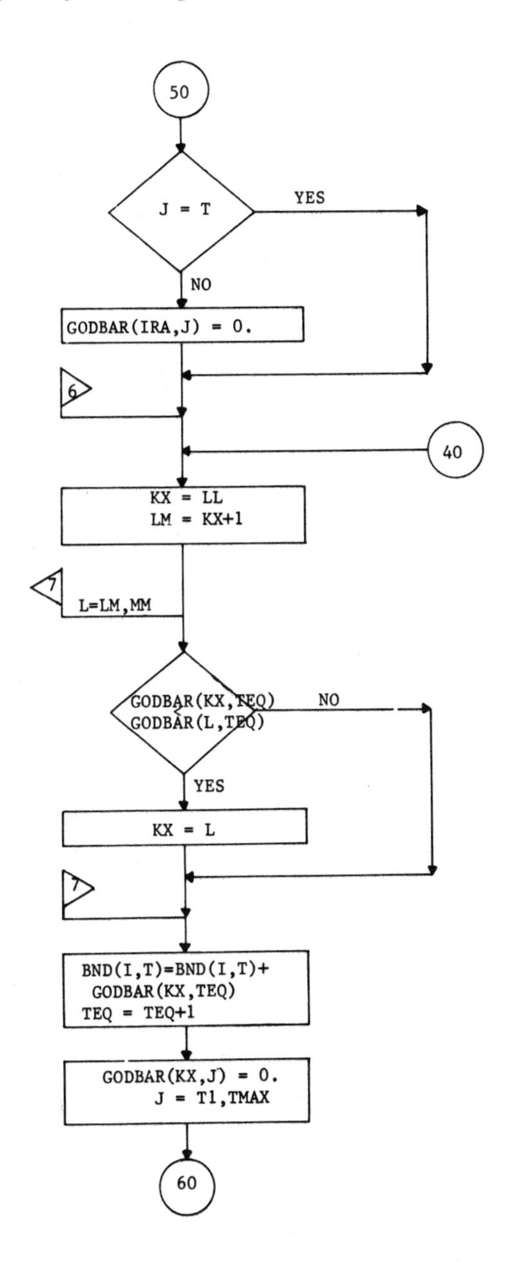
50
J = T
YES
NO
GODBAR(IRA,J) = 0.
6
40
KX = LL
LM = KX+1
7
L=LM,MM
GODBAR(KX,TEQ)
GODBAR(L,TEQ)
NO
YES
KX = L
7
BND(I,T)=BND(I,T)+
GODBAR(KX,TEQ)
TEQ = TEQ+1
GODBAR(KX,J) = 0.
J = T1,TMAX
60

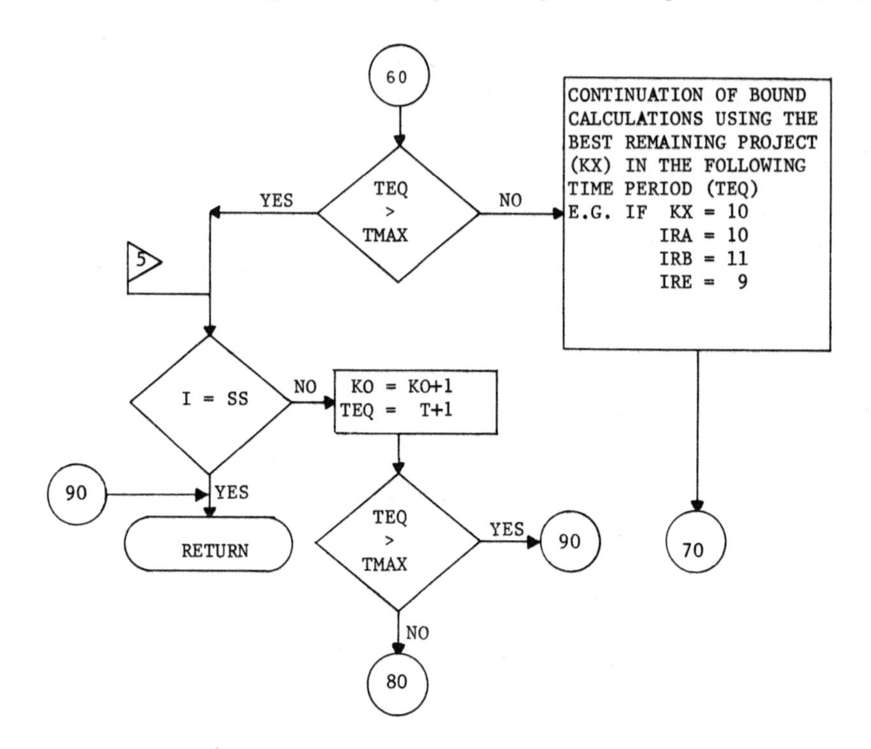
60
TEQ > TMAX
YES
NO
CONTINUATION OF BOUND CALCULATIONS USING THE BEST REMAINING PROJECT (KX) IN THE FOLLOWING TIME PERIOD (TEQ)
E.G. IF KX = 10
IRA = 10
IRB = 11
IRE = 9
5
I = SS
NO
KO = KO+1
TEQ = T+1
90
YES
RETURN
TEQ > TMAX
YES
90
70
NO
80

A.4.4. Subroutine SPEDAT(KX)

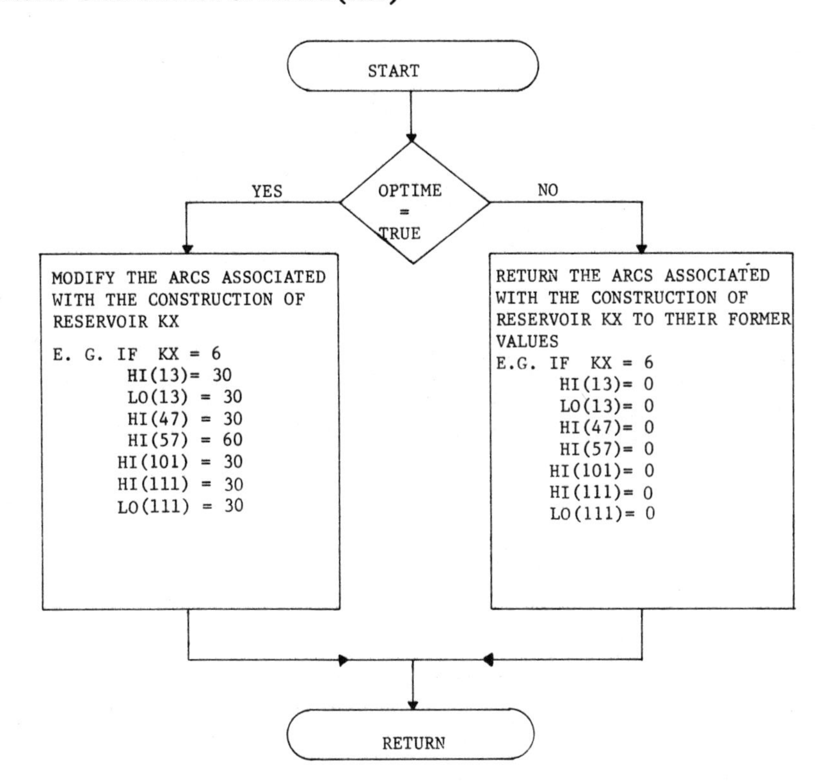

A.4.5. Subroutine READIN

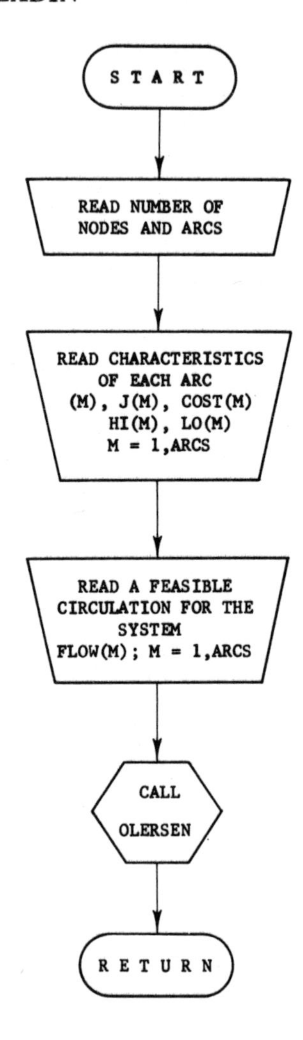

A.4.6. Subroutine OLERSEN

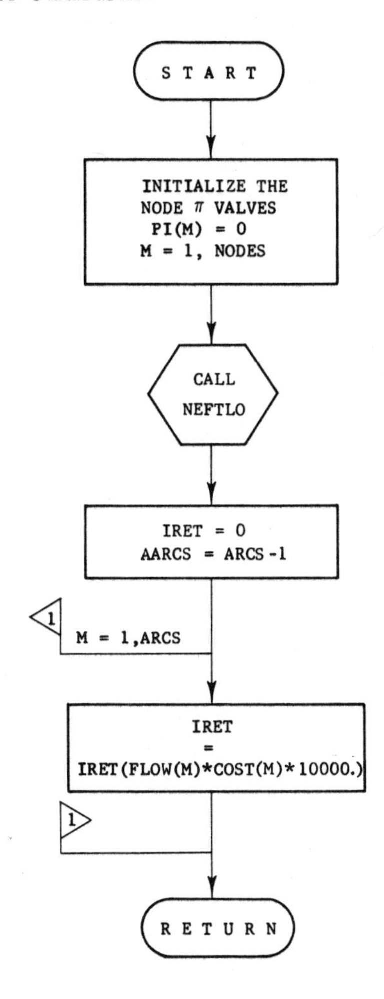

A.4.7. Subroutine PRINTEX

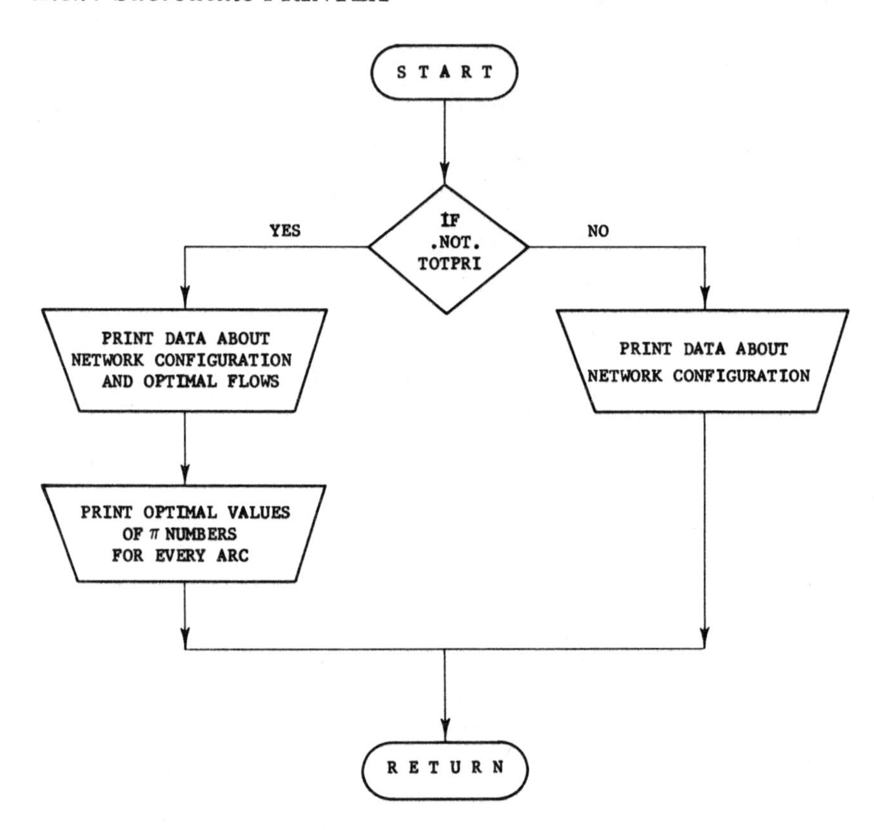

A.4.8. Subroutine EXTDAT

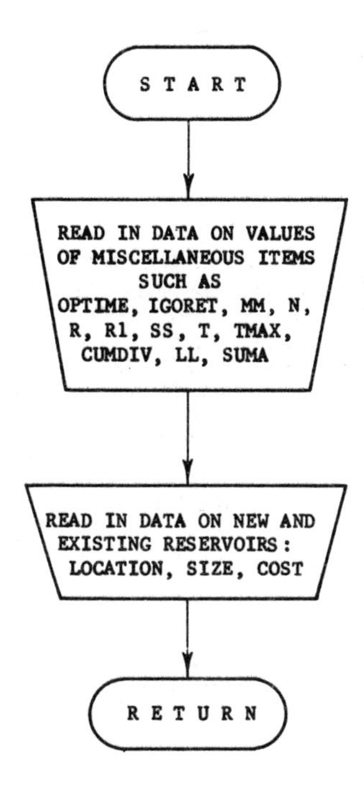

A.4.9. Subroutine KNOKIN(KX)

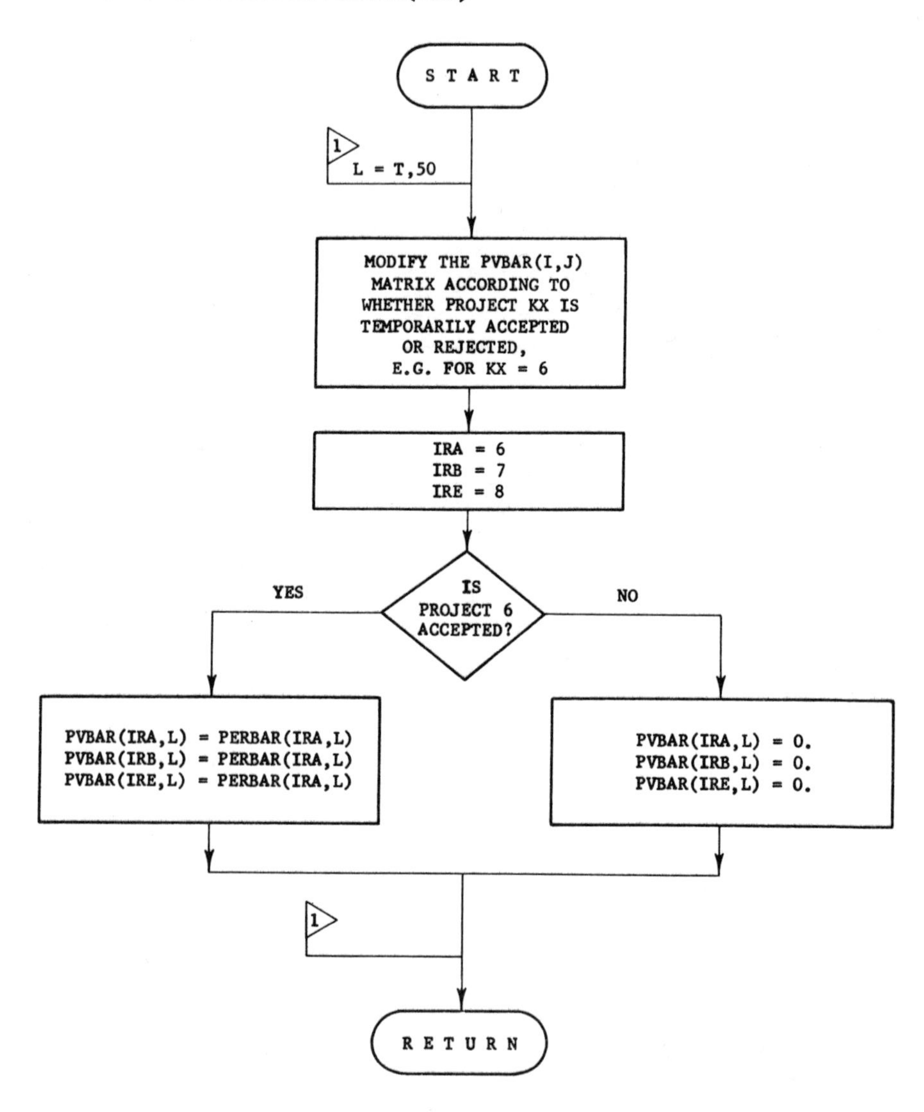

A.4.10. Subroutine NETFLO

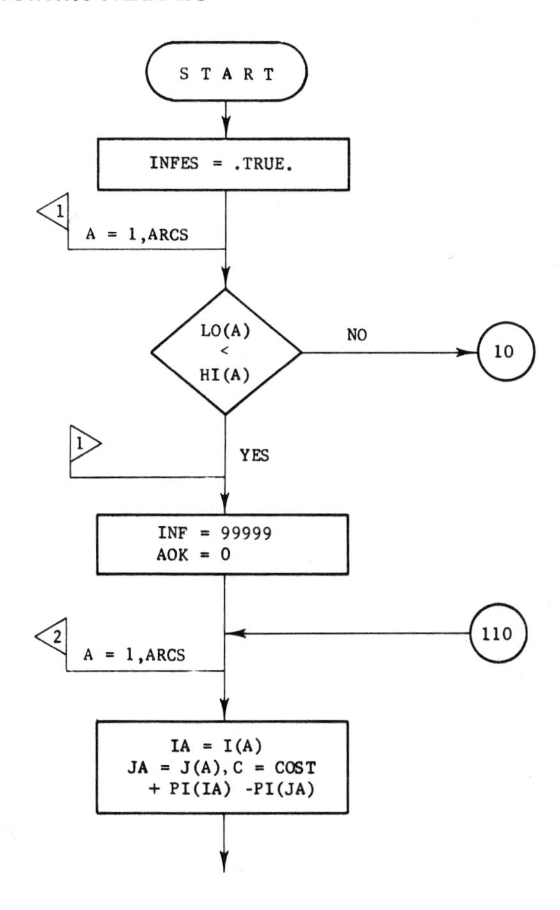

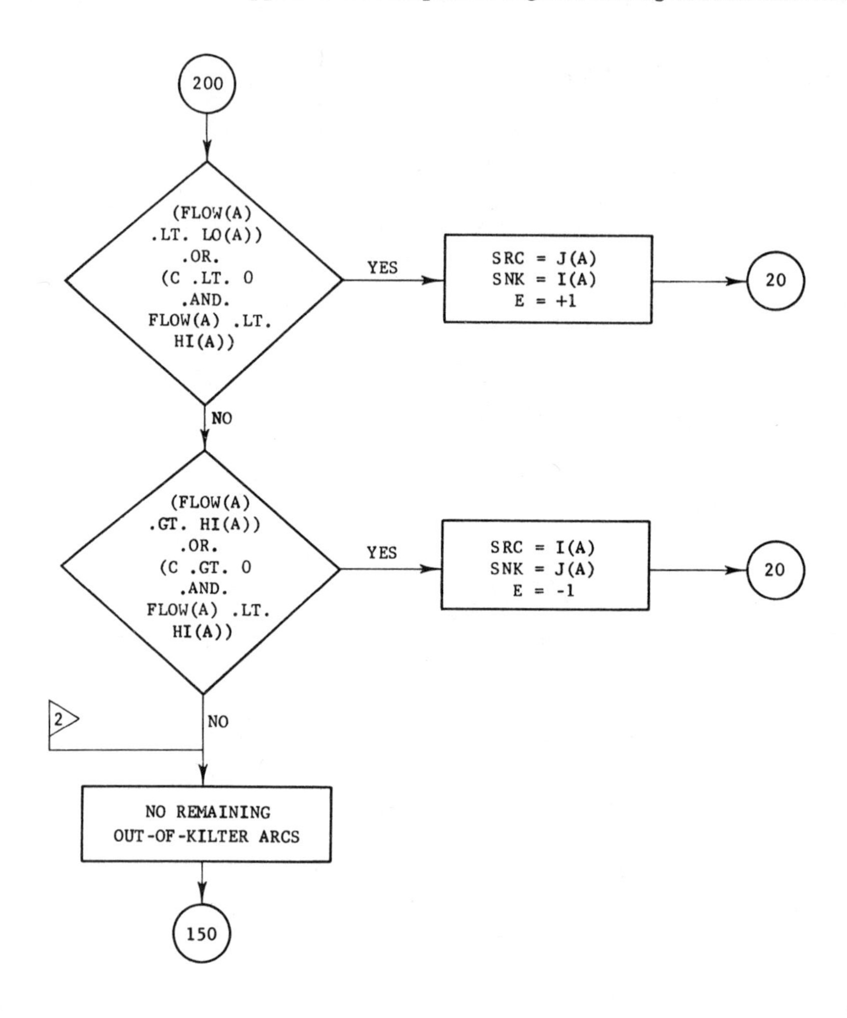
200
(FLOW(A)
.LT. LO(A))
.OR.
(C .LT. 0
.AND.
FLOW(A) .LT.
HI(A))
YES
SRC = J(A)
SNK = I(A)
E = +1
20
NO
(FLOW(A)
.GT. HI(A))
.OR.
(C .GT. 0
.AND.
FLOW(A) .LT.
HI(A))
YES
SRC = I(A)
SNK = J(A)
E = -1
20
2
NO
NO REMAINING
OUT-OF-KILTER ARCS
150

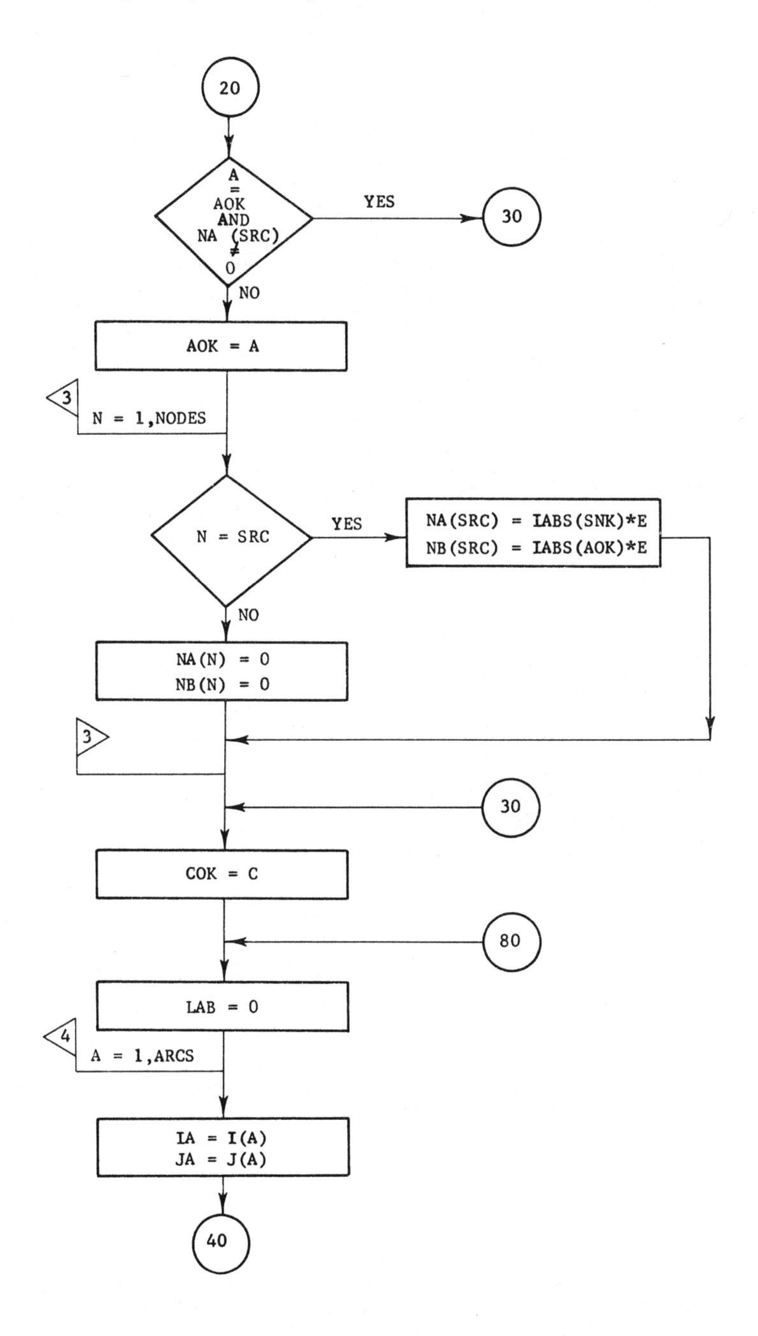
20
A
=
AOK
AND
NA (SRC)
≠
0
YES
30
NO
AOK = A
3
N = 1,NODES
N = SRC
YES
NA(SRC) = IABS(SNK)*E
NB(SRC) = IABS(AOK)*E
NO
NA(N) = 0
NB(N) = 0
3
30
COK = C
80
LAB = 0
4
A = 1,ARCS
IA = I(A)
JA = J(A)
40

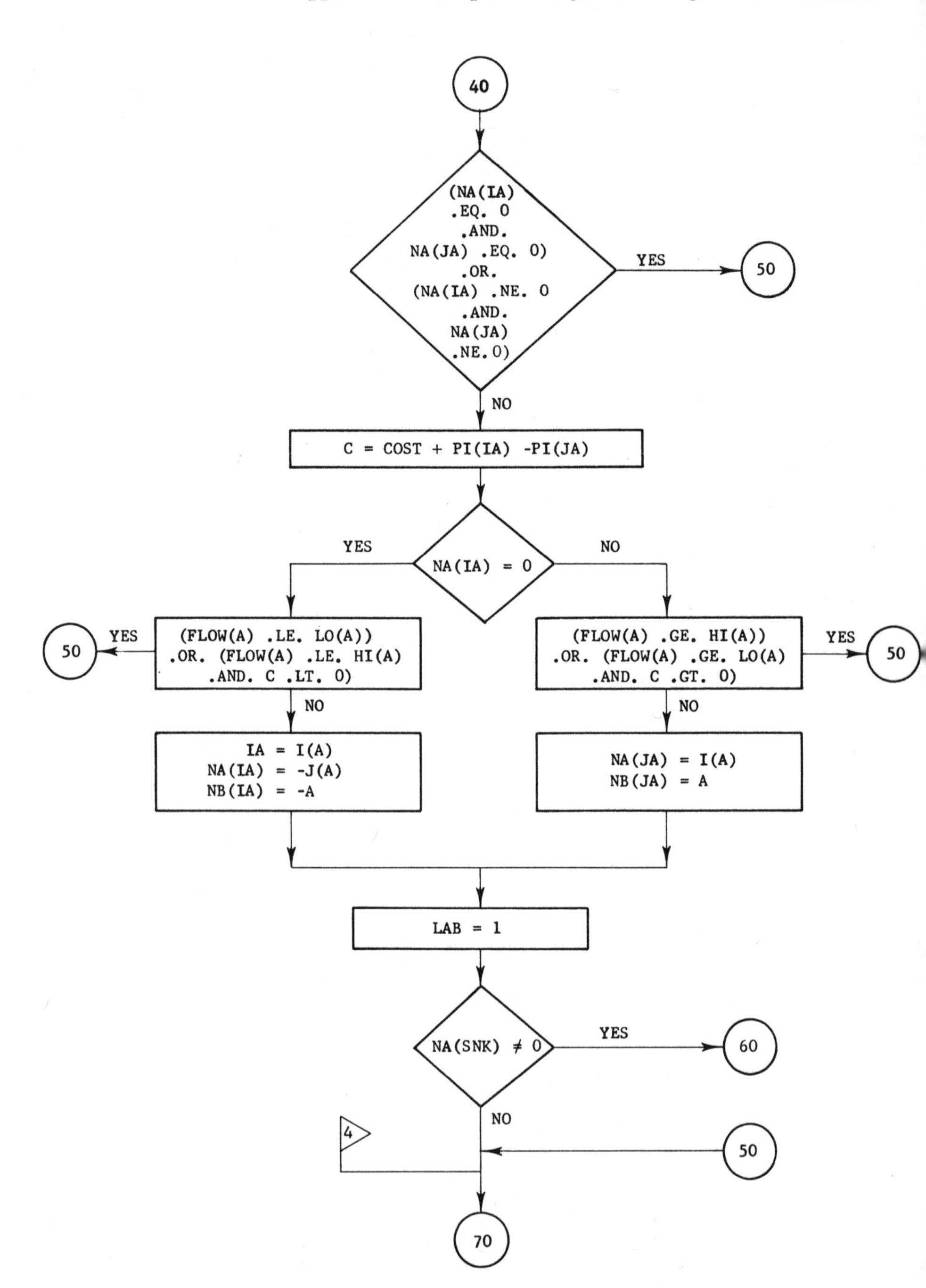
40
(NA(IA)
.EQ. 0
.AND.
NA(JA) .EQ. 0)
.OR.
(NA(IA) .NE. 0
.AND.
NA(JA)
.NE. 0)
YES
50
NO
C = COST + PI(IA) -PI(JA)
YES
NA(IA) = 0
NO
50
YES
(FLOW(A) .LE. LO(A))
.OR. (FLOW(A) .LE. HI(A)
.AND. C .LT. 0)
(FLOW(A) .GE. HI(A))
.OR. (FLOW(A) .GE. LO(A)
.AND. C .GT. 0)
YES
50
NO
NO
IA = I(A)
NA(IA) = -J(A)
NB(IA) = -A
NA(JA) = I(A)
NB(JA) = A
LAB = 1
NA(SNK) ≠ 0
YES
60
NO
4
50
70

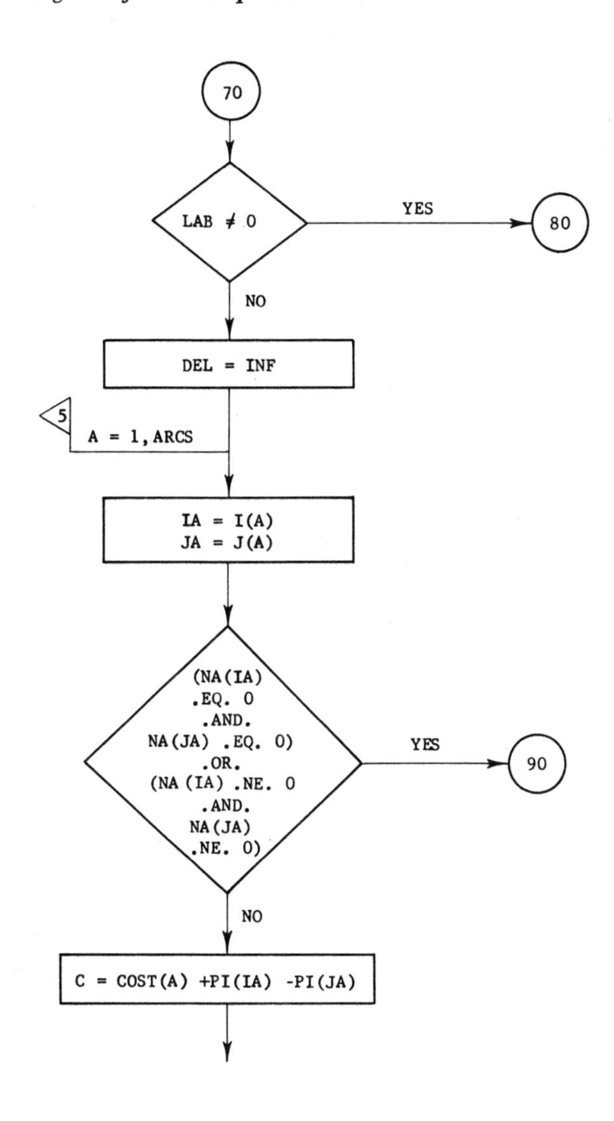
70
LAB ≠ 0
YES
80
NO
DEL = INF
5
A = 1, ARCS
IA = I(A)
JA = J(A)
(NA(IA) .EQ. 0 .AND. NA(JA) .EQ. 0) .OR. (NA(IA) .NE. 0 .AND. NA(JA) .NE. 0)
YES
90
NO
C = COST(A) +PI(IA) -PI(JA)

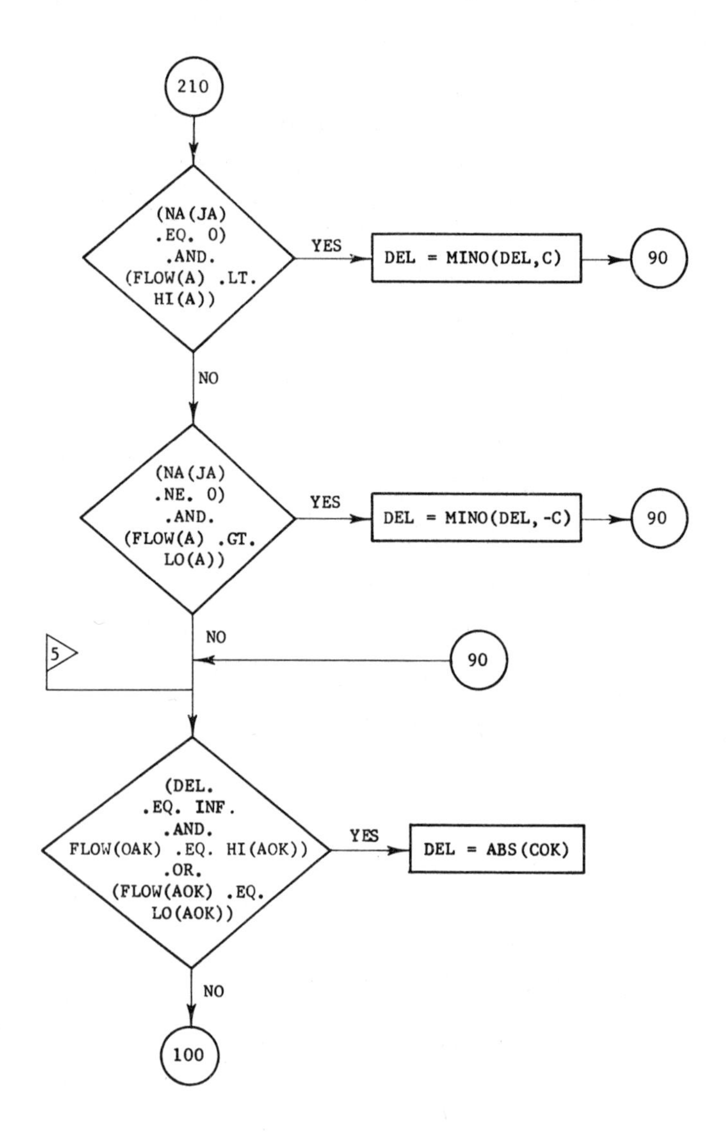
210
(NA(JA)
.EQ. 0)
.AND.
(FLOW(A) .LT.
HI(A))
YES
DEL = MINO(DEL,C)
90
NO
(NA(JA)
.NE. 0)
.AND.
(FLOW(A) .GT.
LO(A))
YES
DEL = MINO(DEL, -C)
90
NO
5
90
(DEL.
.EQ. INF.
.AND.
FLOW(OAK) .EQ. HI(AOK))
.OR.
(FLOW(AOK) .EQ.
LO(AOK))
YES
DEL = ABS(COK)
NO
100

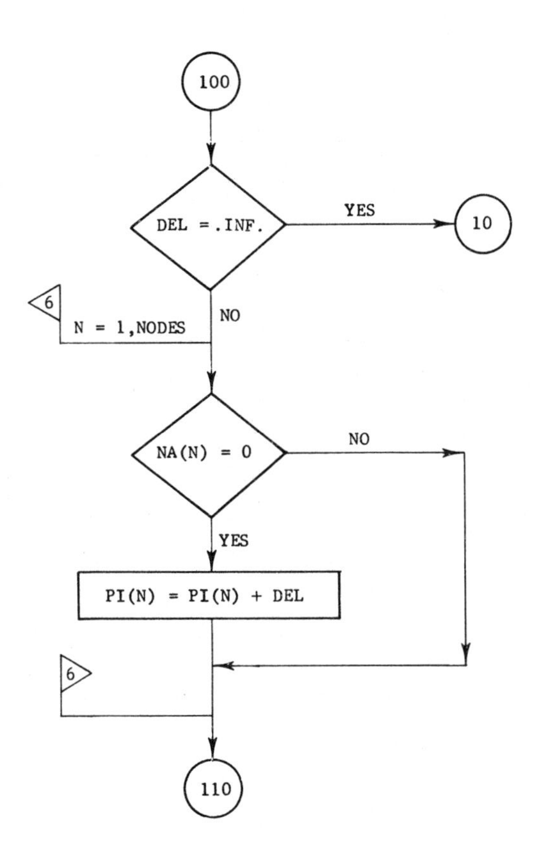
100
DEL = .INF.
YES
10
NO
6
N = 1,NODES
NA(N) = 0
NO
YES
PI(N) = PI(N) + DEL
6
110

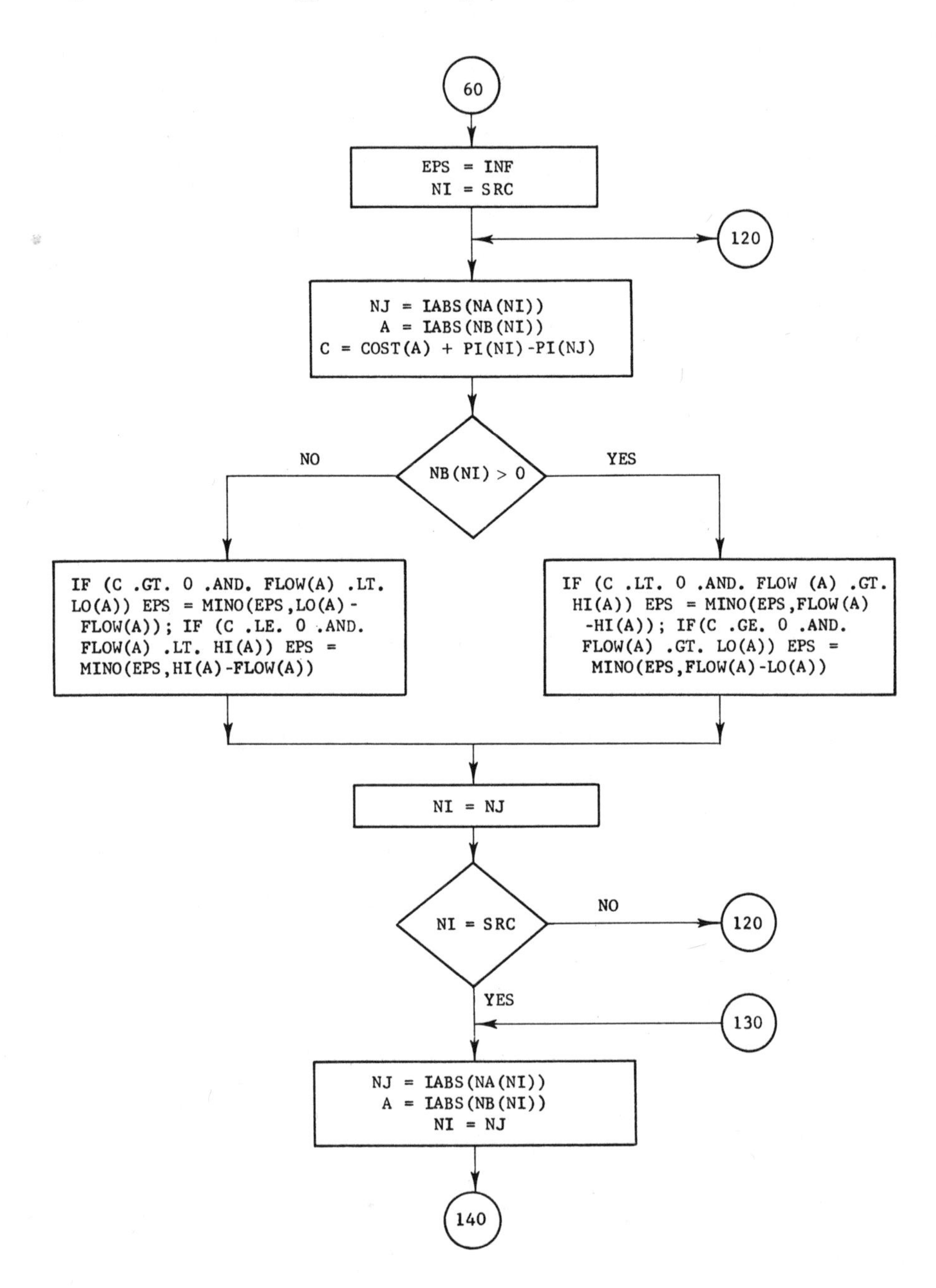

60
EPS = INF
NI = SRC
120
NJ = IABS(NA(NI))
A = IABS(NB(NI))
C = COST(A) + PI(NI)-PI(NJ)
NO
NB(NI) > 0
YES
IF (C .GT. 0 .AND. FLOW(A) .LT. LO(A)) EPS = MINO(EPS,LO(A) - FLOW(A)); IF (C .LE. 0 .AND. FLOW(A) .LT. HI(A)) EPS = MINO(EPS,HI(A)-FLOW(A))
IF (C .LT. 0 .AND. FLOW (A) .GT. HI(A)) EPS = MINO(EPS,FLOW(A) -HI(A)); IF(C .GE. 0 .AND. FLOW(A) .GT. LO(A)) EPS = MINO(EPS,FLOW(A)-LO(A))
NI = NJ
NI = SRC
NO
120
YES
130
NJ = IABS(NA(NI))
A = IABS(NB(NI))
NI = NJ
140

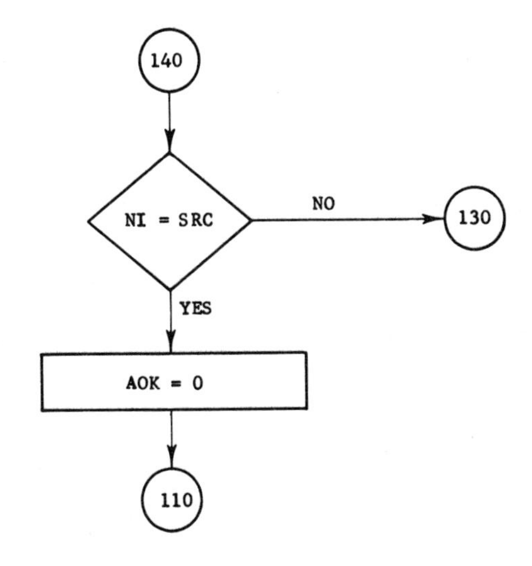
140
NI = SRC
NO
130
YES
AOK = 0
110

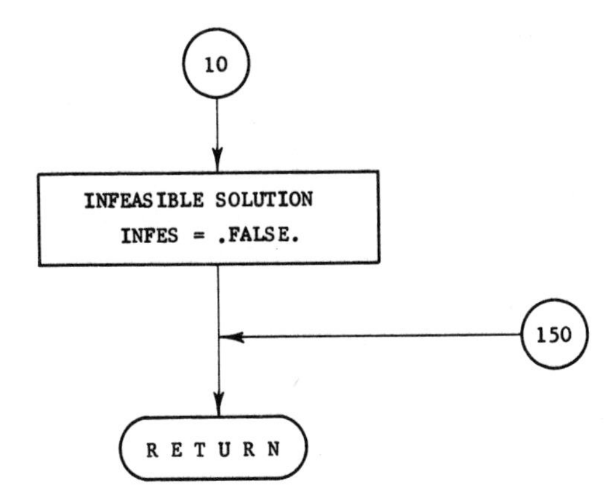
10
INFEASIBLE SOLUTION
INFES = .FALSE.
150
RETURN

A.5. Computer Program Notation

The important variables that appear in routines DAMBLD, BAKTRAK, BOUND, READIN, PRINTEX, EXTDAT, SPEDAT(KX), KNOKIN(KX), OLERSEN, and NETFLO, are defined in this section. The remaining variables correspond to those mentioned in Chapter 3, Section 3.

Three kinds of variables were used in the writing of the computer program: "ordinary," logical, and counting variables. Ordinary variables included fixed-point and floating-point variables; logical variables were used in FORTRAN IV logical statements; and counting variables (with a fixed-point format) were used to enumerate different operations in the program, such as the total number of feasible solutions that had been identified by the program.

The "ordinary" variables are identified in Section A.5.1, the logical variables are listed in Section A.5.2, and the counting variables are detailed in Section A.5.3. Variables that occur only in Subroutine NETFLO are listed in Section A.5.4.

A.5.1. Ordinary Variables

Variable	*Routine(s) Where Used*	*Significance*	*Units*
AARCS	OLERSEN	Temporary variable used in OLERSEN	
AOPCST(K)	DAMBLD, EXTDAT	Annual operating cost for project K	$
ARCS	BAKTRAK, DAMBLD, NETFLO, PRINTEX, OLERSEN, READIN	Total number of arcs	
AVCAP(T)	BAKTRAK, DAMBLD, EXTDAT	Available capital for year T	$
BND(K,T)	BAKTRAK, BOUND, DAMBLD	Bound associated with node K of year T	$
CAPAC(L)	DAMBLD, SPEDAT(KX)	Maximum active storage (i.e., amount above the dead storage level) of reservoir L	acre-ft

Variable	*Routine(s) Where Used*	*Significance*	*Units*
COST(K)	NETFLO, OLERSEN, PRINTEX, READIN	Cost of passing one unit of flow through arc K	$/acre-ft
DEBT	BAKTRAK, DAMBLD, EXTDAT	Original debt	$
DELIM	BAKTRAK, DAMBLD, EXTDAT	Debt limit	$
FLOW(K)	NETFLO, OLERSEN, PRINTEX, READIN	Flow in arc K	acre-ft
HI(K)	BAKTRAK DAMBLD, NETFLO, OLERSEN, PRINTEX, READIN, SPEDAT(KX)	Maximum allowable flow in arc K	acre-ft
I(K)	NETFLO, OLERSEN, PRINTEX, READIN	Source node for arc K	
IGORET(T)	BAKTRAK, BOUND, DAMBLD	Return from operating the system in year T	$
INV(K)	BAKTRAK, BOUND, DAMBLD, EXTDAT	Investment needed for building project K	$
IRET	BAKTRAK, DAMBLD, OLERSEN	Return from operating the system for one year	$
IST(I,J)	BAKTRAK	Element of matrix used for recording the construction sequence of reservoirs e.g. IST(6,2) =1 means build dam 6 in year 2; IST (6,2) =0 means do nothing	

Variable	*Routine(s) Where Used*	*Significance*	*Units*
ISPY(I)	BAKTRAK, NETFLO	Number of non-feasible solutions considered (can be printed out)	
ISUPER	BOUND	Estimates the annual net revenue for the system as it exists at the beginning of the planning horizon; must be calculated by one pass of the program for 1st year	$
J(K)	NETFLO, OLERSEN, PRINTEX, READIN	Sink node for arc K	
KICKS	DAMBLD	Value of HI(54) in time period 2	acre-ft
KORET(T)	DAMBLD, BOUND, BAKTRAK	Return from operating system for year T	$
LIFE(K)	BAKTRAK, BOUND, DAMBLD, EXTDAT	Useful economic life for project K	years
LL	BAKTRAK, DAMBLD, EXTDAT	N + 1	
LO(K)	NETFLO, OLERSEN, PRINTEX, READIN	Minimum allowable flow in arc K	acre-ft
M(I,J)	BAKTRAK, BOUND, DAMBLD	Element of matrix which records the construction sequence of the *best* current solution [M(I,J) = 0 means no construction]	

Variable	*Routine(s) Where Used*	*Significance*	*Units*
MAR(K)	BAKTRAK, BOUND, DAMBLD, EXTDAT	Maximum annual revenue from operating dam K	$
MM	DAMBLD, EXTDAT, PRINTEX	Total of old and new projects	
MS(I)	BOUND	= 1, reservoir I has been included in the construction scheme = 0 otherwise	
N	BAKTRAK, DAMBLD, EXTDAT, PRINTEX	Total number of reservoirs in operation initially	
NODES	BAKTRAK, DAMBLD, OLERSEN, READIN	Total number of nodes	
NONO(K)	DAMBLD, EXTDAT	Node number corresponding to reservoir K	
NRES(K)	EXTDAT, DAMBLD	Index of reservoir K (both old and new)	
PERBAR (I,J)	BAKTRAK, BOUND, DAMBLD, KNOKIN(KX)	Permanent record of PVBAR(I,J)	$
PI(K)	NETFLO, OLERSEN, PRINTEX	Dual variable associated with node K	
PMAX	BAKTRAK, DAMBLD	Net return from the current best solution	$
PVBAR(I,J)	BAKTRAK, BOUND, DAMBLD, KNOKIN(KX)	Net present value from operating project I from year J to the end of the planning horizon TMAX	$

Variable	*Routine(s) Where Used*	*Significance*	*Units*
PVEARN(T)	BAKTRAK, DAMBLD	Present value of net earnings to year T	$
R	BAKTRAK, BOUND, DAMBLD, EXTDAT	Discount rate	
R1	BAKTRAK, BOUND, DAMBLD, EXTDAT	Loan interest rate	
SS	BOUND, DAMBLD	Index used for adding no new projects	
SUMA	DAMBLD, BAKTRAK	Revenue for operating the current configuration for 1 year without addition of new projects	$
TMAX	BOUND, DAMBLD, EXTDAT, KNOKIN(KX), BAKTRAK	Planning horizon	years
TAOPCS	BAKTRAK, DAMBLD, EXTDAT	Total annual operating costs	$

A.5.2. Logical Variables

Variable	*Routine(s) Where Used*	*Significance*
INFES	NETFLO, OLERSEN	= .FALSE. Infeasible solution for network flow problem
OPTIME	BAKTRAK, DAMBLD, EXTDAT, KNOKIN(KX), SPEDAT(KX)	= .TRUE. Change network configuration to add a new dam = .FALSE. Revert network configuration to its previous state
SIMPLE	BAKTRAK	= .FALSE. Used to avoid duplication of "returning parts to their former potency"

Variable	*Routine(s) Where Used*	*Significance*
TOTPRI	DAMBLD, PRINTEX	= .TRUE. Print original network configuration = .FALSE. Print final network configuration
YO	BAKTRAK, BOUND, DAMBLD	= .TRUE. Signifies BOUND is called from DAMBLD = .FALSE. Signifies BOUND is called from BAKTRAK

A.5.3. Counting Variables

Variable	*Routine(s) Where Used*	*Significance*
ISPY	BAKTRAK, NETFLO	Counts nonfeasible solutions
IZ	BAKTRAK, DAMBLD	Identification number of the feasible solutions with return higher than the first feasible solution in a monotonically increasing order
KOPE	BAKTRAK	Identifies the number of feasible solutions, whose bound was higher than the *best current feasible solution* but whose net return was lower than the best current feasible solution
T	BAKTRAK, BOUND	Steps forward in time from a node obtained from a backtrack
TEX	BAKTRAK	Used for backtracking
TOT	BAKTRAK, BOUND, DAMBLD	Equals T − 1
TU	BAKTRAK	Used for going through the IST matrix when returning parts to their "former potency"

A.5.4. Variables That Occur Only in Subroutine NETFLO

Variable	*Significance*	*Units*
AOK	The out-of-kilter arc currently under review	
C	Total cost to the system consumer and distributor of transporting one unit through the arc under review [corresponds to q_{ij} of Eq. (3.16)]	$/acre-ft
COK	Value of C for the out-of-kilter arc currently under review	$/acre-ft
DEL	Amount by which the values of π_i (for $i \in \bar{X}$) are changed	
E	= +1, a "forward" arc is out of kilter = −1, a "reverse" arc is out of kilter	
EPS	Amount by which the flow in each arc in the flow-augmenting path is modified *when breakthrough occurs*	acre-ft.
LAB	= 0, no new *node* has been labeled = 1, at least one new node has been labeled	
NA(N)	Indicates from which *node* node N is labeled	
NB(N)	Indicates though which *arc* node N is labeled	
SNK	Subscript, indicates sink node	
SRC	Subscript, indicates source node	

A.6. A FORTRAN IV Listing of Program DAMBLD

```
                                 PROGRAM DAMBLD(INPUT,OUTPUT)
C                                **********************************            A   2
C**** DIMENSION , COMMON , INTEGER , REAL AND LOGICAL STATEMENTS                A   3
      COMMON /UNOS/ AOPCST(20)                                                  A   4
      COMMON /UNOSA/ HI(300)                                                    A   5
      COMMON /DOS/ NODES, ARCS, INFES, IRET                                     A   6
      COMMON /TRES/ DEBT, DELIM, LL, MM, N, PMAX, R, R1, SS, SUMA, T, TM        A   7
     1AX, TOT                                                                   A   8
      COMMON /CUATRO/ AVCAP(50), CAPAC(20), INV(20), LIFE(20), MAR(20),         A   9
     1NONO(20), NRES(20), PVEARN(50)                                            A  10
      COMMON /CINCO/ OPTIME                                                     A  11
      COMMON /SEIS/ BND(20, 50), IGORET(50), PERBAR(20, 50), PVBAR(20, 5        A  12
     10)                                                                        A  13
      COMMON /SIETE/ M(20, 50)                                                  A  14
      COMMON /OCHO/ IZ, TAOPCS                                                  A  15
      COMMON /NUEVE/ YO, IST(20, 50)                                            A  16
      COMMON /ONCE/ TOTPRI                                                      A  17
      INTEGER HI                                                                A  18
      INTEGER SS                                                                A  19
      INTEGER SUMA                                                              A  20
      INTEGER T, TMAX, T1, TOT                                                  A  21
      LOGICAL OPTIME                                                            A  22
      LOGICAL TOTPRI                                                            A  23
      LOGICAL YO                                                                A  24
      CALL EXTDAT                                                               A  25
      IZ = 1                                                                    A  26
      YO = .TRUE.                                                               A  27
      TOT = 1                                                                   A  28
C**** PRINTING OUTPUT HEADINGS                                                  A  29
      PRINT 450                                                                 A  30
      PRINT 260                                                                 A  31
      PRINT 470                                                                 A  32
      PRINT 540                                                                 A  33
      PRINT 470                                                                 A  34
      PRINT 480                                                                 A  35
      PRINT 550                                                                 A  36
      PRINT 490                                                                 A  37
      PRINT 560                                                                 A  38
      PRINT 570, (NRES(K), NONO(K), CAPAC(K), AOPCST(K), K = 1, N)              A  39
      PRINT 580                                                                 A  40
      PRINT 490                                                                 A  41
      PRINT 590                                                                 A  42
      PRINT 600, (NRES(K), NONO(K), CAPAC(K), INV(K), AOPCST(K), MAR(K),        A  43
     1 LIFE(K), K = LL, MM)                                                     A  44
      CALL READIN                                                               A  45
C**** INITIALISING THE TIMING MATRIX                                            A  46
      DO 10 T = 1, 50                                                           A  47
      DO 10 KX = LL, MM                                                         A  48
 10   M(KX, T) = 0                                                              A  49
C**** CALCULATION OF PRESENT VALUE FOR EACH PROJECT AND YEAR                    A  50
      PRINT 450                                                                 A  51
      PRINT 270                                                                 A  52
      PRINT 280                                                                 A  53
      PRINT 290                                                                 A  54


      PRINT 300                                                                 A  55
      DO 30 T = 1, 50                                                           A  56
      DO 20 I = LL, MM                                                          A  57
      AA = 1/((1+R)**T)                                                         A  58
      AB = 1/((1+R)**(T+LIFE(I)))                                               A  59
```

```
      AC = AA-AB                                                      A  60
      AD = MAR(I)*AC                                                  A  61
      AD = AD/R                                                       A  62
      AE = INV(I)*AA                                                  A  63
      PVBAR(I, T) = AD-AE                                             A  64
      IF (PVBAR(I, T) .GE. 1.0) GO TO 20                              A  65
      PVBAR(I, T) = 0.0                                               A  66
 20   CONTINUE                                                        A  67
 30   CONTINUE                                                        A  68
      DO 40 J = 1, 50                                                 A  69
      DO 40 I = LL, MM                                                A  70
      PERBAR(I, J) = PVBAR(I, J)                                      A  71
 40   CONTINUE                                                        A  72
      PRINT 640, (((PVBAR(I, J)), I = LL, MM), J = 1, TMAX)           A  73
      PRINT 450                                                       A  74
      PRINT 310                                                       A  75
      PRINT 320                                                       A  76
      TOTPRI = .TRUE.                                                 A  77
      CALL PRINTEX                                                    A  78
      T = 1                                                           A  79
50    IF (T.LE.TMAX) GO TO 60
      PRINT 450                                                       A  81
      PRINT 610, T                                                    A  82
      PRINT 490                                                       A  83
      PRINT 630, AVCAP(T)                                             A  84
 60   T1 = T+1                                                        A  85
C**** FINDING CURRENT OPERATING COSTS                                  A  86
      TAOPCS = 0.                                                     A  87
      DO 70 K = LL, MM                                                A  88
      TAOPCS = TAOPCS+AOPCST(K)                                       A  89
 70   CONTINUE                                                        A  90
      OPTIME = .TRUE.                                                 A  91
C**** CALLING THE NETWORK ALGORITHM                                    A  92
      IF (T .GE. 30) GO TO 80                                         A  93
      CALL OLERSEN                                                    A  94
      IF (T .NE. 1) GO TO 80                                          A  95
      PRINT 450                                                       A  96
      PRINT 330                                                       A  97
      PRINT 340                                                       A  98
      TOTPRI = .FALSE.                                                A  99
      CALL PRINTEX                                                    A 100
      PRINT 450                                                       A 101
 80   SUMA = IRET                                                     A 102
      CALL BOUND                                                      A 103
 90   KX = LL                                                         A 104
      LM = KX+1                                                       A 105
      DO 100 I = LM, MM                                               A 106
      IF (BND(KX, T) .LT. BND(I, T)) KX = I                           A 107
 100  CONTINUE                                                        A 108

      IF (BND(KX, T) .LT. 0.1) GO TO 150                              A 109
      IF (INV(KX) .LE. AVCAP(T)) GO TO 110                            A 110
      IF (INV(KX) .GT. AVCAP(T)) BND(KX, T) = 0.                      A 111
      GO TO 90                                                        A 112
 110  CALL SPEDAT (KX)                                                A 113
      CALL OLERSEN                                                    A 114
      IF (IRET .LE. SUMA) GO TO 140                                   A 115
      M(KX, T) = 1                                                    A 116
      BND(KX, T) = 0.                                                 A 117
      OPTIME = .FALSE.                                                A 118
      CALL KNOKIN (KX)                                                A 119
```

```
      IF (T .GT. 1) GO TO 120                                                  A 120
      PVEARN(1) = IRET/((1+R)**T)-TAOPCS/((1+R)**T)-INV(KX)/((1+R)**T)         A 121
      GO TO 130                                                                A 122
 120  PVEARN(T) = PVEARN(TOT)+IRET/((1+R)**T)-TAOPCS/((1+R)**T)-INV(KX)/       A 123
     1((1+R)**T)                                                               A 124
C**** CALCULATION OF ANNUAL BUDGETARY CONSTRAINT                               A 125
 130  AVCAP(T1) = AVCAP(T)+((IRET-TAOPCS)/((1+R)**T))-INV(KX)                  A 126
      GO TO 180                                                                A 127
 140  OPTIME = .FALSE.                                                         A 128
      CALL SPEDAT (KX)                                                         A 129
 150  IRET = SUMA                                                              A 130
      BND(SS, T) = 0.                                                          A 131
      IF (T .GT. 1) GO TO 160                                                  A 132
      PVEARN(1) = IRET/((1+R)**T)-TAOPCS/((1+R)**T)                            A 133
      GO TO 170                                                                A 134
 160  PVEARN(T) = PVEARN(TOT)+IRET/((1+R)**T)-TAOPCS/((1+R)**T)                A 135
C**** CALCULATION OF ANNUAL BUDGETARY CONSTRAINT                               A 136
 170  AVCAP(T1) = AVCAP(T)+((IRET-TAOPCS)/((1+R)**T))                          A 137
 180  T = T+1                                                                  A 138
      TOT = T-1                                                                A 139
      IGORET(TOT) = IRET                                                       A 140
      HI(52) = HI(52)+2                                                        A 141
      HI(54) = HI(54)+2                                                        A 142
      IF (T .EQ. 2) KICKS = HI(54)                                             A 143
      IF (HI(52) .GT. 27) HI(52) = 30                                          A 144
      IF (HI(54) .GT. 55) HI(54) = 60                                          A 145
      IF (T .GT. TMAX) GO TO 190                                               A 146
      GO TO 50                                                                 A 147
 190  CONTINUE                                                                 A 148
      PMAX = PVEARN(TMAX)                                                      A 149
      PRINT 450                                                                A 150
      PRINT 350                                                                A 151
      PRINT 360                                                                A 152
      PRINT 370, IZ                                                            A 153
      PRINT 380                                                                A 154
      DO 200 J = 1, TMAX                                                       A 155
      DO 200 I = LL, MM                                                        A 156
      IF (M(I, J) .EQ. 1) PRINT 440, I, J                                      A 157
 200  CONTINUE                                                                 A 158
      PRINT 670, PVEARN(TMAX)                                                  A 159
      PRINT 450                                                                A 160
      PRINT 660                                                                A 161
      PRINT 500                                                                A 162


      PRINT 390                                                                A 163
      PRINT 650, (((BND(I, J)), I = LL, SS), J = 1, 50)                        A 164
      PRINT 450                                                                A 165
      CALL BAKTRAK                                                             A 166
C**** SUMMARY OF OUTPUT RESULTS                                                A 167
      PRINT 450                                                                A 168
      PRINT 400                                                                A 169
      PRINT 510                                                                A 170
      PRINT 710                                                                A 171
      PRINT 510                                                                A 172
      PRINT 450                                                                A 173
      PRINT 660                                                                A 174
      PRINT 500                                                                A 175
      PRINT 390                                                                A 176
      PRINT 650, (((BND(I, J)), I = LL, SS), J = 1, 50)                        A 177
      PRINT 620                                                                A 178
      PRINT 450                                                                A 179
```

```
      PRINT 460                                                          A 180
      PRINT 530                                                          A 181
      DO 210 J = 1, TMAX                                                 A 182
      DO 210 I = LL, MM                                                  A 183
      IF (M(I, J) .EQ. 1) PRINT 440, I, J                                A 184
  210 CONTINUE                                                           A 185
      PRINT 680                                                          A 186
      PRINT 520                                                          A 187
      PRINT 690, PMAX                                                    A 188
      PRINT 700, TMAX                                                    A 189
      PRINT 450
      PRINT 410                                                          A 190
      PRINT 420                                                          A 191
  220 DO 230 T = 1, TMAX                                                 A 192
      DO 230 I = LL, MM                                                  A 193
      IF (M(I, T) .NE. 1.) GO TO 230                                     A 194
      IF (M(I, T) .EQ. 1) IS = I                                         A 195
      M(I, T) = 0                                                        A 196
      GO TO 240                                                          A 197
  230 CONTINUE                                                           A 198
      GO TO 250                                                          A 199
  240 OPTIME = .TRUE.                                                    A 200
      CALL SPEDAT (IS)                                                   A 201
      GO TO 220                                                          A 202
  250 CALL OLERSEN                                                       A 203
      TOTPRT = .FALSE.                                                   A 204
      CALL PRINTEX                                                       A 205
      PRINT 450                                                          A 206
      PRINT 430, R, KICKS                                                A 207
      RETURN                                                             A 209
C**** FORMAT STATEMENTS                                                  A 208
C                                                                        A 210
  260 FORMAT (10/)                                                       A 211
  270 FORMAT (   58X,*THE PVBAR MATRIX*)                                 A 212
  280 FORMAT (55X,20(1H*))                                               A 213
  290 FORMAT (2/,10X,* COLUMNS CORRESPOND TO NEW RESERVOIRS(6 TO 14)*)   A 214
  300 FORMAT (10X,* ROWS CORRESPOND TO YEARS(1 TO 50)*,2/)               A 215
  310 FORMAT (5/,29X,*THE INITIAL NETWORK CONFIGURATION*)                A 216


  320 FORMAT (26X,39(1H*),2/)                                            A 217
  330 FORMAT (5/,16X,*THE OPERATING POLICY FOR THE ORIGINAL NETWORK CONF A 218
     1IGURATION*)                                                        A 219
  340 FORMAT (13X,64(1H*),//)                                            A 220
  350 FORMAT (2/,19X,*INTERMEDIATE RESULTS*)                             A 221
  360 FORMAT (19X,26(1H*))                                               A 222
  370 FORMAT (4/,16X,*FEASIBLE SOLUTION NUMBER*,I4)                      A 223
  380 FORMAT (13X,35(1H*),2/)                                            A 224
  390 FORMAT (2/,10X,*COLUMNS(1 TO  9) CORRESPOND TO NEW RESERVOIRS(6 TO A 225
     1 14)*,/,10X,*COLUMN 10 CORRESPONDS TO NO NEW RESERVOIR*,/,10X,*ROW A 226
     2S CORRESPOND TO YEARS(1 TO 50)*,2/)                                A 227
  400 FORMAT (26/)                                                       A 228
  410 FORMAT (4/,28X,*THE FINAL NETWORK CONFIGURATION*)                  A 229
  420 FORMAT (25X,37(1H*),2/)                                            A 230
  430 FORMAT (2X,*INTEREST RATE =*,F12.6,2X,*HI(54) FOR (T =2) =*,I7)    A 231
  440 FORMAT    (13X,*DAM*,I3,* WAS CONSTRUCTED AND BEGAN OPERATION IN YE A 232
     1AR*,I3)                                                            A 233
  450 FORMAT     (1H1)                                                   A 234
  460 FORMAT    (5/,29X,*TIMING OF CONSTRUCTION OF DAMS*)                A 235
  470 FORMAT    (45X,44(1H*))                                            A 236
  480 FORMAT    (5/)                                                     A 237
  490 FORMAT    (7X,25(1H*))                                             A 238
```

```
500  FORMAT   (55X,22(1H*))                                                     A 239
510  FORMAT ( (34X,23(1H*))                                                     A 240
520  FORMAT   (35X,20(1H*))                                                     A 241
530  FORMAT   (29X,36(1H*),2/)                                                  A 242
540  FORMAT   (51X,*OPTIMISATION OF THE EXPANSION OF*,/,50X,*AN EXISTIN          A 243
    1G WATER RESOURCES SYSTEM*)                                                 A 244
550  FORMAT   (10X,*EXISTING RESERVOIRS*)                                       A 245
560  FORMAT   (29X,*RESERVOIR          REPRESENTED            CAPACITY          A 246
    1            ANNUAL*,/32X,*NUMBER            BY   NODE              (ACR    A 247
    2E-FT.)        OPERATING COSTS*)                                            A 248
570  FORMAT   (32X,I5,14X,I5,09X,F10.0,12X,F10.0)                               A 249
580  FORMAT   (4/,15X,*NEW RESERVOIRS*)                                         A 250
590  FORMAT   ( 1X,*RFSERVOIR          REPRESENTED            CAPACITY          A 251
    1          REQUIRED               ANNUAL                MAX.ANNUAL          A 252
    2           USEFUL*,/, 3X,*NUMBER          BY      NODE           (A        A 253
    3CRE-FT.)           INVESTMENT         OPERATING COSTS            R         A 254
    4ETURN               LIFE(YRS.)*)                                           A 255
600  FORMAT   ( 1X,I5,15X,I5,14X,F10.0,10X,I10,10X,F10.0,13X,I10,13X,I          A 256
    15)                                                                         A 257
610  FORMAT   (2/,7X,*FINANCIAL DATA FOR YEAR*,I3)                              A 258
    1  AXIS*)                                                                   A 259
620  FORMAT   (2/,4X,*SINCE ALL THE ELEMENTS ARE EQUAL TO ZERO,WE THUS          A 260
    1HAVE SEARCHED THROUGH EVERY BRANCH OF THE TREE*)                           A 261
630  FORMAT   (/,4X,*AVAILABLE CAPITAL =*,F10.0,*DOLLARS*)                      A 262
640  FORMAT (1X,8(F10.0,5X),F10.0)                                              A 263
650  FORMAT (1X,9(F10.0,3X),F10.0)                                              A 264
660  FORMAT (3/,58X,*THE BOUND MATRIX*)                                         A 265
670  FORMAT (2/,13X,*THE TOTAL PRESENT VALUE OF EARNINGS =*,F10.0,*DOLL         A 266
    1ARS*)                                                                      A 267
680  FORMAT   (5/,38X,*TOTAL EARNINGS*)                                         A 268
690  FORMAT   (2/,18X,*THE TOTAL PRESENT VALUE OF NETT EARNINGS =*,F10.0        A 269
    1,*DOLLARS*)                                                                A 270


700  FORMAT   (2/,18X,*THE SYSTEM WAS OPERATED FOR*,I3,*YEARS*)                 A 271
710  FORMAT ( (37X,*FINAL RESULTS*)                                             A 272
     END                                                                        A 273-
```

```
                              SUBROUTINE BAKTRAK
C                          *************************              B   2
C**** DIMENSION , COMMON , INTEGER , REAL AND LOGICAL STATEMENTS   B   3
      COMMON /UNOSA/ HI(300)                                      B   4
      COMMON /DOS/ NODES, ARCS, INFES, IRET                       B   5
      COMMON /TRES/ DEBT, DELIM, LL, MM, N, PMAX, R, R1, SS, SUMA, T, TM B   6
     1AX, TOT                                                     B   7
      COMMON /CUATRO/ AVCAP(50), CAPAC(20), INV(20), LIFE(20), MAR(20), B   8
     1NONO(20), NRES(20), PVFARN(50)                              B   9
      COMMON /CINCO/ OPTIME                                       B  10
      COMMON /SEIS/ BND(20, 50), IGORET(50), PERBAR(20, 50), PVBAR(20, 5 B  11
     10)                                                          B  12
      COMMON /SIETE/ M(20, 50)                                    B  13
      COMMON /OCHO/ IZ, TAOPCS                                    B  14
      COMMON /NUEVE/ YO, IST(20, 50)                              B  15
      INTEGER HI                                                  B  16
      INTEGER SS                                                  B  17
      INTEGER SUMA                                                B  18
      INTEGER T, TMAX, TOT, T1                                    B  19
      INTEGER TEX, TU                                             B  20
      LOGICAL OPTIME                                              B  21
      LOGICAL SIMPLE                                              B  22
      LOGICAL YO                                                  B  23
C**** CREATE A TEMPORARY FILE ON NEW PROJECTS                      B  24
      YO = .FALSE.                                                B  25
      KOPE = 0                                                    B  26
      DO 10 J = 1, TMAX                                           B  27
      DO 10 I = LL, MM                                            B  28
      IST(I, J) = M(I, J)                                         B  29
 10   CONTINUE                                                    B  30
      T = TMAX                                                    B  31
      TOT = T-1                                                   B  32
      TEX = TMAX                                                  B  33
      TU = TEX                                                    B  34
      SIMPLE = .TRUE.                                             B  35
C**** CHOOSING THE LARGEST BOUND                                   B  36
 20   T1 = T+1                                                    B  37
 30   KX = LL                                                     B  38
      LM = KX+1                                                   B  39
      DO 40 I = LM, SS                                            B  40
      IF (BND(KX, T) .LT. BND(I, T)) KX = I                       B  41
 40   CONTINUE                                                    B  42
      IF (T .EQ. TEX) KXI = KX                                    B  43
C**** IS THIS BOUND GREATER THAN ZERO OR LARGER THAN PMAX          B  44
      IF (TEX .NE. T) GO TO 50                                    B  45
      IF (BND(KX, TEX) .LT. 0.1) GO TO 220                        B  46
      IF (BND(KX, TEX) .LE. PMAX) GO TO 60                        B  47
 50   IF (KX .EQ. SS) GO TO 80                                    B  48
      IF (BND(KX, T) .LT. 0.1) GO TO 120                          B  49
      IF (INV(KX) .LE. AVCAP(T)) GO TO 80                         B  50
      IF (INV(KX) .GT. AVCAP(T)) BND(KX, T) = 0.0                 B  51
      GO TO 30                                                    B  52
C**** THE CASE WHEN LARGEST BOUND IS LESS THAN PMAX                B  53
C**** LOOK AT ALTERNATIVE WITH BOUND HIGHER THAN PMAX              B  54


 60   DO 70 I = LL, SS                                            B  55
      BND(I, TEX) = 0.                                            B  56
 70   CONTINUE                                                    B  57
      GO TO 220                                                   B  58
C**** REMOVE CONFIGURATION ADDED IN YEAR T I.E. PICK OUT THE CONFIGURAT- B  59
```

```
C**** ION ADDED IN YEAR T AND SUBSEQUENT YEARS                          B  60
 80   IF ( .NOT. SIMPLE) GO TO 120                                      B  61
 90   IS = 0                                                            B  62
      DO 100 I = LL, MM                                                 B  63
      IF (IST(I, TU) .EQ. 0) GO TO 100                                  B  64
      IF (IST(I, TU) .EQ. 1) IS = I                                     B  65
      IST(I, TU) = 0                                                    B  66
 100  CONTINUE                                                          B  67
      IF (IS .EQ. 0) GO TO 110                                          B  68
      OPTIME = .FALSE.                                                  B  69
      CALL SPEDAT (IS)                                                  B  70
C**** RETURN KNOCKED OUT PARTS TO THEIR FORMER POTENCY                   B  71
      OPTIME = .TRUE.                                                   B  72
      CALL KNOKIN (IS)                                                  B  73
 110  TU = TU+1                                                         B  74
      IF (TU .LE. TMAX) GO TO 90                                        B  75
      SIMPLE = .FALSE.                                                  B  76
C**** CALCULATE RETURN BEFORE NEW PARTS ARE ADDED AND MODIFY SOME ARCS   B  77
 120  HI(52) = 15+2*(T-1)                                               B  78
      HI(54) = 35+2*(T-1)                                               B  79
      IF (HI(52) .GT. 27) HI(52) = 30                                   B  80
      IF (HI(54) .GT. 55) HI(54) = 60                                   B  81
      IF (T .GE. 20) GO TO 130                                          B  82
      CALL OLERSEN                                                      B  83
 130  SUMA = IRET                                                       B  84
C     ADDITION OF NEW PARTS                                             B  85
      IF (KX .EQ. SS) GO TO 160                                         B  86
      IF (BND(KX, T) .LT. 0.1) GO TO 160                                B  87
      OPTIME = .TRUE.                                                   B  88
      CALL SPEDAT (KX)                                                  B  89
      CALL OLERSEN                                                      B  90
C**** CRITERION FOR ADDING KX (WE THINK) TEMPORARILY                     B  91
      IF (IRET .LE. SUMA) GO TO 150                                     B  92
      DO 140 IS = 1, SS                                                 B  93
      IF (IS .EQ. KX) IST(IS, T) = 1                                    B  94
      IF (IS .NE. KX) IST(IS, T) = 0                                    B  95
 140  CONTINUE                                                          B  96
      BND(KX, T) = 0.                                                   B  97
      IF (T .EQ. 1) PVEARN(1) = (IRET-TAOPCS)/((1+R)**T)-INV(KX)/((1+R)* B  98
     1*T)                                                               B  99
      PVEARN(T) = PVEARN(TOT)+IRET/((1+R)**T)-TAOPCS/((1+R)**T)-INV(KX)/ B 100
     1((1+R)**T)                                                        B 101
C**** CALCULATION OF BUDGETARY CONSTRAINT                                B 102
      AVCAP(T1) = AVCAP(T)+((IRET-TAOPCS)/((1+R)**T))-INV(KX)           B 103
      OPTIME = .FALSE.                                                  B 104
      CALL KNOKIN (KX)                                                  B 105
      GO TO 180                                                         B 106
C**** REVERTING TO OLD SCENE                                             B 107
C**** NO CHANGE IN IST                                                   B 108

 150  OPTIME = .FALSE.                                                  B 109
      CALL SPEDAT (KX)                                                  B 110
 160  DO 170 IS = 1, SS                                                 B 111
      IST(IS, T) = 0                                                    B 112
 170  CONTINUE                                                          B 113
      BND(KX, T) = 0.                                                   B 114
      IRET = SUMA                                                       B 115
      IF (T .EQ. 1) PVEARN(1) = (IRET-TAOPCS)/((1+R)**T)                B 116
      PVEARN(T) = PVEARN(TOT)+IRET/((1+R)**T)-TAOPCS/((1+R)**T)         B 117
C**** CALCULATION OF BUDGETARY CONSTRAINT                                B 118
      AVCAP(T1) = AVCAP(T)+((IRET-TAOPCS)/((1+R)**T))                   B 119
```

```
      GO TO 180                                                          B 120
 180  T = T+1                                                            B 121
      IF (T .GT. 50) GO TO 190                                           B 122
C**** PREPARE FOR THE FOLLOWING YEAR                                     B 123
      TOT = T-1                                                          B 124
      IGORET(TOT) = IRET                                                 B 125
      CALL BOUND                                                         B 126
      GO TO 20                                                           B 127
C**** TEST TO SEE IF NEW PVEARN(TMAX) IS GREATER THAN PMAX               B 128
 190  IF (PVEARN(TMAX) .LE. PMAX) GO TO 210                              B 129
      IF (PVEARN(TMAX) .GT. PMAX) PMAX = PVEARN(TMAX)                    B 130
      IZ = IZ+1                                                          B 131
      PRINT 270, IZ                                                      B 132
      PRINT 280                                                          B 133
      DO 200 J = 1, TMAX                                                 B 134
      DO 200 I = LL, MM                                                  B 135
      M(I, J) = IST(I, J)                                                B 136
      IF (M(I, J) .EQ. 0) GO TO 200                                      B 137
      PRINT 290, I, J                                                    B 138
 200  CONTINUE                                                           B 139
      PRINT 300, PVEARN(TMAX)                                            B 140
      BND(KXI, TEX) = 0.                                                 B 141
      T = T-1                                                            B 142
      TU = T                                                             B 143
      TEX = T                                                            B 144
      TOT = T-1                                                          B 145
      SIMPLE = .TRUE.                                                    B 146
      GO TO 20                                                           B 147
C**** WHAT HAPPENS WHEN PVEARN(TMAX) IS LESS THAN PMAX                   B 148
 210  BND(KXI, TEX) = 0.                                                 B 149
      KOPE = KOPE+1                                                      B 150
      T = T-1                                                            B 151
      TU = T                                                             B 152
      TEX = T                                                            B 153
      TOT = T-1                                                          B 154
      SIMPLE = .TRUE.                                                    B 155
      GO TO 20                                                           B 156
 220  TEX = TEX-1                                                        B 157
      T = TEX                                                            B 158
      TU = T                                                             B 159
      TOT = T-1                                                          B 160
      SIMPLE = .TRUE.                                                    B 161
      IF (TEX .EQ. 1) GO TO 230                                          B 162


      GO TO 20                                                           B 163
C     MAKING THE FINAL NETWORK CONFIGURATION CORRESPOND TO THE OPTIMUM C  B 164
C     ONFIGURATION                                                       B 165
 230  DO 240 T = 1, TMAX                                                 B 166
      DO 240 I = LL, MM                                                  B 167
      IF (IST(I, T) .NE. 1) GO TO 240                                    B 168
      IF (IST(I, T) .EQ. 1) IS = I                                       B 169
      IST(I, T) = 0                                                      B 170
      GO TO 250                                                          B 171
 240  CONTINUE                                                           B 172
      GO TO 260                                                          B 173
 250  OPTIME = .FALSE.                                                   B 174
      CALL SPEDAT (IS)                                                   B 175
      GO TO 230                                                          B 176
 260  CONTINUE                                                           B 177
      RETURN                                                             B 178
C**** FORMAT STATEMENTS
C                                                                        B 179
 270  FORMAT (4/,16X,*FEASIBLE SOLUTION NUMBER*,I4)                      B 180
 280  FORMAT (13X,35(1H*),2/)                                            B 181
 290  FORMAT   (13X,*DAM*,I3,* WAS CONSTRUCTED AND BEGAN OPERATION IN YE  B 182
     1AR*,I3)                                                            B 183
 300  FORMAT (2/,13X,*THE TOTAL PRESENT VALUE OF EARNINGS =*,F10.0,*DOLL  B 184
     1ARS*)                                                              B 185
      END                                                                B 186-
```

```
                        SUBROUTINE BOUND
C                       ********************                              C   2
C**** DIMENSION , COMMON , INTEGER , REAL AND LOGICAL STATEMENTS           C   3
      DIMENSION GODBAR(20,50), ISUPGO(50), MS(20)                          C   4
      COMMON /TRES/ DEBT, DELIM, LL, MM, N, PMAX, R, R1, SS, SUMA, T, TM   C   5
     1AX, TOT                                                              C   6
      COMMON /CUATRO/ AVCAP(50), CAPAC(20), INV(20), LIFE(20), MAR(20),    C   7
     1NONO(20), NRES(20), PVEARN(50)                                       C   8
      COMMON /SEIS/ BND(20, 50), IGORET(50), PERBAR(20, 50), PVBAR(20, 5   C   9
     10)                                                                   C  10
      COMMON /SIETE/ M(20, 50)                                             C  11
      COMMON /NUEVE/ YO, IST(20, 50)                                       C  12
      INTEGER SS                                                           C  13
      INTEGER T, TMAX                                                      C  14
      INTEGER T1, TEQ, TOT                                                 C  15
      LOGICAL YO                                                           C  16
      KO = LL                                                              C  17
      TEQ = T+1                                                            C  18
      T1 = T+1                                                             C  19
C**** CALCULATE THE RETURN FROM OPERATING THE PRESENT SYSTEM TO THE END    C  20
C**** OF THE PLANNING HORIZON                                              C  21
      DO 10 I = LL, MM                                                     C  22
      MS(I) = 0                                                            C  23
 10   CONTINUE                                                             C  24
      IF ( .NOT. YO) GO TO 30                                              C  25
      DO 20 I = LL, MM                                                     C  26
      DO 20 J = 1, T                                                       C  27
      IF (M(I, J) .EQ. 1) MS(I) = 1                                        C  28
 20   CONTINUE                                                             C  29
      GO TO 50                                                             C  30
 30   DO 40 I = LL, MM                                                     C  31
      DO 40 J = 1, T                                                       C  32
      IF (IST(I, J) .EQ. 1) MS(I) = 1                                      C  33
 40   CONTINUE                                                             C  34
 50   ISUPGO(TOT) = IGORET(TOT)                                            C  35
      ISUPER=7350000
      DO 60 I = LL, MM                                                     C  37
      IF (MS(I) .EQ. 1) ISUPER = ISUPER+MAR(I)                             C  38
 60   CONTINUE                                                             C  39
      ISUPGO(TOT) = MAX0(ISUPGO(TOT), ISUPER)                              C  40
      AA = 1/((1+R)**(T-1))                                                C  41
      AB = 1/((1+R)**50)                                                   C  42
      AC = AA-AB                                                           C  43
      AD = ISUPGO(TOT)*AC                                                  C  44
      AF=ISUPER*AC
      BIGO = AD/R                                                          C  46
      BIGOS = AF/R                                                         C  47
C**** INITIALIZE THE BOUND VARIABLES                                       C  48
      DO 100 I = LL, SS                                                    C  49
      IF (I .EQ. SS .AND. T .EQ. 1) GO TO 80                               C  50
      IF (I .EQ. SS .AND. T .GT. 1) GO TO 90                               C  51
      IF (PVBAR(I, T) .LT. 0.1) BND(I, T) = 0.0                            C  52
      IF (PVBAR(I, T) .LT. 0.1) GO TO 100                                  C  53
      IF (T .GT. 1) GO TO 70                                               C  54


      BND(I, 1) = PVBAR(I, 1)+BIGOS                                        C  55
      GO TO 100                                                            C  56
 70   BND(I, T) = PVBAR(I, T)+PVEARN(TOT)+BIGO                             C  57
      GO TO 100                                                            C  58
 80   BND(SS, T) = BIGOS                                                   C  59
```

```
      GO TO 100                                                          C  60
90    BND(SS, T) = BIGO+PVEARN(TOT)                                      C  61
100   CONTINUE                                                           C  62
110   DO 120 I = LL, MM                                                  C  63
      DO 120 J = 1, 50                                                   C  64
120   GODBAR(I, J) = PVBAR(I, J)                                         C  65
C**** CALCULATION OF BOUNDS FOR ALL VARIABLES                            C  66
      DO 270 I = KO, SS                                                  C  67
      IF (I .EQ. SS) GO TO 240                                           C  68
      IF (BND(I, T) .LT. 0.1) GO TO 280                                  C  69
      IF (I .EQ. 6) GO TO 130                                            C  70
      IF (I .EQ. 7) GO TO 140                                            C  71
      IF (I .EQ. 8) GO TO 150                                            C  72
      IF (I .EQ. 9) GO TO 160                                            C  73
      IF (I .EQ. 10) GO TO 170                                           C  74
      IF (I .EQ. 11) GO TO 180                                           C  75
      IF (I .EQ. 12) GO TO 190                                           C  76
      IF (I .EQ. 13) GO TO 200                                           C  77
      IF (I .EQ. 14) GO TO 210                                           C  78
130   IRA = 6                                                            C  79
      IRB = 7                                                            C  80
      IRE = 8                                                            C  81
      GO TO 220                                                          C  82
140   IRA = 7                                                            C  83
      IRB = 8                                                            C  84
      IRE = 6                                                            C  85
      GO TO 220                                                          C  86
150   IRA = 8                                                            C  87
      IRB = 6                                                            C  88
      IRE = 7                                                            C  89
      GO TO 220                                                          C  90
160   IRA = 9                                                            C  91
      IRB = 10                                                           C  92
      IRE = 11                                                           C  93
      GO TO 220                                                          C  94
170   IRA = 10                                                           C  95
      IRB = 11                                                           C  96
      IRE = 9                                                            C  97
      GO TO 220                                                          C  98
180   IRA = 11                                                           C  99
      IRB = 9                                                            C 100
      IRE = 10                                                           C 101
      GO TO 220                                                          C 102
190   IRA = 12                                                           C 103
      IRB = 13                                                           C 104
      IRE = 14                                                           C 105
      GO TO 220                                                          C 106
200   IRA = 13                                                           C 107
      IRB = 14                                                           C 108


      IRE = 12                                                           C 109
      GO TO 220                                                          C 110
210   IRA = 14                                                           C 111
      IRB = 12                                                           C 112
      IRE = 13                                                           C 113
      GO TO 220                                                          C 114
C**** MAKING THE RETURN EQUAL TO ZERO FOR I (AND ALL PROJECTS EXCLUSIVE   C 115
C**** TO I) AND FOR ALL YEARS AFTER YEAR T                               C 116
220   DO 230 J = T, 50                                                   C 117
      GODBAR(IRB, J) = 0.                                                C 118
      GODBAR(IRE, J) = 0.                                                C 119
```

```
      IF (J .EQ. T) GO TO 230                                          C 120
      GODBAR(IRA, J) = 0.                                              C 121
 230  CONTINUE                                                         C 122
 240  KX = LL                                                          C 123
      LM = KX+1                                                        C 124
      DO 250 L = LM, MM                                                C 125
      IF (GODBAR(KX, TEQ) .LT. GODBAR(L, TEQ)) KX = L                  C 126
 250  CONTINUE                                                         C 127
      BND(I, T) = BND(I, T)+GODBAR(KX, TEQ)                            C 128
      DO 260 J = T1, TMAX                                              C 129
      GODBAR(KX, J) = 0.                                               C 130
 260  CONTINUE                                                         C 131
      TEQ = TEQ+1                                                      C 132
      IF (TEQ .GT. TMAX) GO TO 280                                     C 133
      IF (KX .EQ. 6) GO TO 130                                         C 134
      IF (KX .EQ. 7) GO TO 140                                         C 135
      IF (KX .EQ. 8) GO TO 150                                         C 136
      IF (KX .EQ. 9) GO TO 160                                         C 137
      IF (KX .EQ. 10) GO TO 170                                        C 138
      IF (KX .EQ. 11) GO TO 180                                        C 139
      IF (KX .EQ. 12) GO TO 190                                        C 140
      IF (KX .EQ. 13) GO TO 200                                        C 141
      IF (KX .EQ. 14) GO TO 210                                        C 142
 270  CONTINUE                                                         C 143
 280  IF (I .EQ. SS) GO TO 290                                         C 144
      KO = KO+1                                                        C 145
      TEQ = T+1                                                        C 146
      IF (TEQ .GT. TMAX) GO TO 290
      GO TO 110                                                        C 147
 290  CONTINUE                                                         C 148
      RETURN                                                           C 149
      END                                                              C 150-
```

```
                         SUBROUTINE READIN
C                        ************************          R          D   2
C**** DIMENSION , COMMON , INTEGER , REAL AND LOGICAL STATEMENTS        D   3
      COMMON /UNO/ I(250), J(250), COST(250), LO(250), FLOW(250), PI(250 D   4
     1)                                                                 D   5
      COMMON /UNOSA/ HI(300)                                            D   6
      COMMON /DOS/ NODES, ARCS, INFES, IRET                             D   7
      INTEGER ARCS, AARCS, COST, FLOW                                   D   8
      INTEGER HI                                                        D   9
      READ 10, NODES, ARCS                                              D  10
      READ 20, (I(M), J(M), M = 1, ARCS)                                D  11
      READ 30, (COST(M), M = 1, ARCS)                                   D  12
      READ 30, (HI(M), M = 1, ARCS)                                     D  13
      READ 30, (LO(M), M = 1, ARCS)                                     D  14
      READ 30, (FLOW(M), M = 1, ARCS)                                   D  15
C**** FORMAT STATEMENTS                                                 D  16
      RETURN                                                            D  17
C                                                                       D  18
 10   FORMAT     (2I5)                                                  D  19
 20   FORMAT     (8(I6,I4))                                             D  20
 30   FORMAT     (8I10)                                                 D  21
      END                                                               D  22-
```

```
                      SUBROUTINE PRINTEX
C                    **********************                          E   2
C**** DIMENSION , COMMON , INTEGER , REAL AND LOGICAL STATEMENTS     E   3
      COMMON /UNO/ I(250), J(250), COST(250), LO(250), FLOW(250), PI(250 E   4
     1)                                                              E   5
      COMMON /UNOSA/ HI(300)                                         E   6
      COMMON /DOS/ NODES, ARCS, INFES, IRET                          E   7
      COMMON /ONCE/ TOTPRI                                           E   8
      INTEGER ARCS, AARCS, COST, FLOW                                E   9
      INTEGER HI                                                     E  10
      LOGICAL TOTPRI                                                 E  11
      IF ( .NOT. TOTPRI) GO TO 10                                    E  12
      PRINT 30, NODES, ARCS, (M, I(M), J(M), COST(M), HI(M), LO(M), M =  E  13
     11, ARCS)                                                       E  14
      GO TO 20                                                       E  15
 10   PRINT 40, NODES, ARCS, (M, I(M), J(M), COST(M), HI(M), LO(M), FLOW E  16
     1(M), M = 1, ARCS)                                              E  17
      PRINT 60                                                       E  18
      PRINT 50, (M, M = 1, 14)                                       E  19
      PRINT 70, (PI(M), M = 1, 14)                                   E  20
      PRINT 50, (M, M = 15, 28)                                      E  21
      PRINT 70, (PI(M), M = 15, 28)                                  E  22
      PRINT 50, (M, M = 29, 42)                                      E  23
      PRINT 70, (PI(M), M = 29, 42)                                  E  24
      PRINT 50, (M, M = 43, 56)                                      E  25
      PRINT 70, (PI(M), M = 43, 56)                                  E  26
      PRINT 50, (M, M = 57, 70)                                      E  27
      PRINT 70, (PI(M), M = 57, 70)                                  E  28
      PRINT 50, (M, M = 71, 84)                                      E  29
      PRINT 70, (PI(M), M = 71, 84)                                  E  30
      PRINT 50, (M, M = 85, 98)                                      E  31
      PRINT 70, (PI(M), M = 85, 98)                                  E  32
      PRINT 50, (M, M = 99, 112)                                     E  33
      PRINT 70, (PI(M), M = 99, 112)                                 E  34
      PRINT 50, (M, M = 113, 126)                                    E  35
      PRINT 70, (PI(M), M = 113, 126)                                E  36
      PRINT 50, (M, M = 127, 131)                                    E  37
      PRINT 70, (PI(M), M = 127, 131)                                E  38
C**** FORMAT STATEMENTS                                              E  39
 20   RETURN                                                         E  40
C                                                                    E  41
 30   FORMAT (14X,*NUMBER OF NODES = *,I4,/,14X, *NUMBER OF ARCS  =*,I5, E  42
     13/,21X,*M*,5X,*I*,5X,*J*,11X,*COST*, 9X,*HI*,11X,*LO*,//,(16X,3(2X E  43
     2,I4),3(3X,I10)))                                               E  44
 40   FORMAT (14X,*NUMBER OF NODES = *,I4,/,14X, *NUMBER OF ARCS  =*,I5, E  45
     13/,13X,*M*,5X,*I*,5X,*J*,11X,*COST*, 9X,*HI*,10X,*LO*,10X,*FLOW*,/ E  46
     2/,(8X,3(2X,I4),4(3X,I10)))                                     E  47
 50   FORMAT (13X,*      M *,14(I4))                                 E  48
 60   FORMAT (8/)                                                    E  49
 70   FORMAT (2/,14X,*PYE(M)*,14(I4),4/)                             E  50
      END                                                            E  51-
```

```
                    SUBROUTINE SPEDAT(KX)
C                   ****************************                    G    2
C**** DIMENSION , COMMON , INTEGER , REAL AND LOGICAL STATEMENTS     G    3
      COMMON /UNO/ I(250), J(250), COST(250), LO(250), FLOW(250), PI(250 G    4
     1)                                                              G    5
      COMMON /UNOSA/ HI(300)                                         G    6
      COMMON /CINCO/ OPTIME                                          G    7
      INTEGER HI                                                     G    8
      INTEGER FLOW                                                   G    9
      LOGICAL OPTIME                                                 G   10
      IF (KX .EQ. 6) GO TO 10                                        G   11
      IF (KX .EQ. 7) GO TO 30                                        G   12
      IF (KX .EQ. 8) GO TO 50                                        G   13
      IF (KX .EQ. 9) GO TO 70                                        G   14
      IF (KX .EQ. 10) GO TO 90                                       G   15
      IF (KX .EQ. 11) GO TO 110                                      G   16
      IF (KX .EQ. 12) GO TO 130                                      G   17
      IF (KX .EQ. 13) GO TO 150                                      G   18
      IF (KX .EQ. 14) GO TO 170                                      G   19
 10   IF ( .NOT. OPTIME) GO TO 20                                    G   20
      HI(13) = 30                                                    G   21
      LO(13) = 30                                                    G   22
      HI(47) = 30                                                    G   23
      HI(57) = 60                                                    G   24
      HI(101) = 30                                                   G   25
      HI(111)=30
      LO(111)=30
      GO TO 190                                                      G   27
 20   HI(13) = 0                                                     G   28
      LO(13) = 0                                                     G   29
      HI(47) = 0                                                     G   30
      HI(57) = 0                                                     G   31
      HI(101) = 0                                                    G   32
      HI(111) = 0                                                    G   33
      LO(111)=0
      GO TO 190                                                      G   34
 30   IF ( .NOT. OPTIME) GO TO 40                                    G   35
      HI(13) = 25                                                    G   36
      LO(13) = 25                                                    G   37
      HI(47) = 27                                                    G   38
      HI(57) = 50                                                    G   39
      HI(101) = 27                                                   G   40
      HI(111)=25
      LO(111)=25
      GO TO 190                                                      G   42
 40   HI(13) = 0                                                     G   43
      LO(13) = 0                                                     G   44
      HI(47) = 0                                                     G   45
      HI(57) = 0                                                     G   46
      HI(101) = 0                                                    G   47
      HI(111) = 0                                                    G   48
      LO(111)=0
      GO TO 190                                                      G   49
 50   IF ( .NOT. OPTIME) GO TO 60                                    G   50
      LO(13) = 20                                                    G   51
      HI(13) = 20                                                    G   52
      HI(47) = 24                                                    G   53
      HI(57) = 40                                                    G   54

      HI(101) = 24                                                   G   55
```

```
      HI(111)=20
      LO(111)=20
      GO TO 190                                                        G  58
60    HI(13) = 0                                                       G  59
      LO(13) = 0                                                       G  60
      HI(47) = 0                                                       G  61
      HI(57) = 0                                                       G  62
      HI(101) = 0                                                      G  63
      HI(111) = 0                                                      G  64
      LO(111)=0
      GO TO 190                                                        G  65
70    IF ( .NOT. OPTIMF) GO TO 80                                      G  66
      HI(15) = 12                                                      G  67
      LO(15) = 12                                                      G  68
      HI(59) = 12                                                      G  69
      HI(113) = 12                                                     G  70
      LO(113)=12
      GO TO 190                                                        G  71
80    HI(15) = 0                                                       G  72
      LO(15) = 0                                                       G  73
      HI(59) = 0                                                       G  74
      HI(113) = 0                                                      G  75
      LO(113)=0
      GO TO 190                                                        G  76
90    IF ( .NOT. OPTIMF) GO TO 100                                     G  77
      HI(15) = 10                                                      G  78
      LO(15) = 10                                                      G  79
      HI(59) = 10                                                      G  80
      HI(113) = 10                                                     G  81
      LO(113)=10
      GO TO 190                                                        G  82
100   HI(15) = 0                                                       G  83
      LO(15) = 0                                                       G  84
      HI(59) = 0                                                       G  85
      HI(113) = 0                                                      G  86
      LO(113)=0
      GO TO 190                                                        G  87
110   IF ( .NOT. OPTIMF) GO TO 120                                     G  88
      HI(15) = 8                                                       G  89
      LO(15) = 8                                                       G  90
      HI(59) = 8                                                       G  91
      HI(113) = 8                                                      G  92
      LO(113)=8
      GO TO 190                                                        G  93
120   HI(15) = 0                                                       G  94
      LO(15) = 0                                                       G  95
      HI(59) = 0                                                       G  96
      HI(113) = 0                                                      G  97
      LO(113)=0
      GO TO 190                                                        G  98
130   IF ( .NOT. OPTIME) GO TO 140                                     G  99
      HI(17) = 20                                                      G 100
      LO(17) = 20                                                      G 101
      HI(61) = 20                                                      G 102
      HI(115) = 20                                                     G 103
      LO(115)=20
      GO TO 190                                                        G 104
140   HI(17) = 0                                                       G 105
      LO(17) = 0                                                       G 106
      HI(61) = 0                                                       G 107
      HI(115) = 0                                                      G 108
      LO(115)=0
```

```
      GO TO 190                                                          G 109
150   IF ( .NOT. OPTIME) GO TO 160                                       G 110
      HI(17) = 16                                                        G 111
      LO(17) = 16                                                        G 112
      HI(61) =16
      HI(115)=16
      LO(115)=16
      GO TO 190                                                          G 115
160   HI(17) = 0                                                         G 116
      LO(17) = 0                                                         G 117
      HI(61) = 0                                                         G 118
      HI(115) = 0                                                        G 119
      LO(115)=0
      GO TO 190                                                          G 120
170   IF ( .NOT. OPTIME) GO TO 180                                       G 121
      HI(17) = 10                                                        G 122
      LO(17) = 10                                                        G 123
      HI(61) = 10                                                        G 124
      HI(115) = 10                                                       G 125
      LO(115)=10
      GO TO 190                                                          G 126
180   HI(17) = 0                                                         G 127
      LO(17) = 0                                                         G 128
      HI(61) = 0                                                         G 129
      HI(115) = 0                                                        G 130
      LO(115)=0
      GO TO 190                                                          G 131
190   CONTINUE                                                           G 132
      RETURN                                                             G 133
      END                                                                G 134-
```

```
                                  SUBROUTINE KNOKIN(KX)
C                           ***********************                    H    2
      COMMON /TRES/ DEBT, DELIM, LL, MM, N, PMAX, R, R1, SS, SUMA, T, TM  H    3
     1AX, TOT                                                           H    4
      COMMON /CINCO/ OPTIME                                             H    5
      COMMON /SEIS/ BND(20, 50), IGORET(50), PERBAR(20, 50), PVBAR(20, 5 H    6
     10)                                                                H    7
      LOGICAL OPTIME                                                    H    8
      INTEGER T, TMAX, T1, TOT
      T1 = T+1
      DO 60 L = T, 50
      IF(KX.EQ. 6) GO TO 10
      IF(KX.EQ. 7) GO TO 10
      IF (KX .EQ. 8) GO TO 10                                           H   14
      IF (KX .EQ. 9) GO TO 20                                           H   15
      IF (KX .EQ. 10) GO TO 20                                          H   16
      IF (KX .EQ. 11) GO TO 20                                          H   17
      IF (KX .EQ. 12) GO TO 30                                          H   18
      IF (KX .EQ. 13) GO TO 30                                          H   19
      IF (KX .EQ. 14) GO TO 30                                          H   20
 10   IRA = 6                                                           H   21
      IRB = 7                                                           H   22
      IRE = 8                                                           H   23
      GO TO 40                                                          H   24
 20   IRA = 9                                                           H   25
      IRB = 10                                                          H   26
      IRE = 11                                                          H   27
      GO TO 40                                                          H   28
 30   IRA = 12                                                          H   29
      IRB = 13                                                          H   30
      IRE = 14                                                          H   31
      GO TO 40                                                          H   32
 40   IF ( .NOT. OPTIME) GO TO 50                                       H   33
      PVBAR(IRA, L) = PERBAR(IRA, L)                                    H   34
      PVBAR(IRB, L) = PERBAR(IRB, L)                                    H   35
      PVBAR(IRE, L) = PERBAR(IRE. L)                                    H   36
      GO TO 60                                                          H   37
 50   PVBAR(IRA, L) = 0.                                                H   38
      PVBAR(IRB, L) = 0.                                                H   39
      PVBAR(IRE, L) = 0.                                                H   40
 60   CONTINUE                                                          H   41
      RETURN                                                            H   42
      END                                                               H   43-
```

```
                          SUBROUTINE -OLERSEN
C                      ************************                         I    2
C**** DIMENSION , COMMON , INTEGER , REAL AND LOGICAL STATEMENTS        I    3
      COMMON /UNO/ I(250), J(250), COST(250), LO(250), FLOW(250), PI(250 I    4
     1)                                                                 I    5
      COMMON /UNOSA/ HI(300)                                            I    6
      COMMON /DOS/ NODES, ARCS, INFES, IRET                             I    7
      COMMON /TRES/ DEBT, DELIM, LL, MM, N, PMAX, R, R1, SS, SUMA, T, TM I    8
     1AX, TOT                                                           I    9
      INTEGER ARCS, AARCS, COST, FLOW                                   I   10
      INTEGER HI                                                        I   11
      LOGICAL INFES                                                     I   12
      DO 10 M = 1, NODES                                                I   13
 10   PI(M) = 0                                                         I   14
      CALL NETFLO                                                       I   15
      AARCS = ARCS-1                                                    I   17
      IRET = 0                                                          I   18
      DO 20 M = 1, AARCS                                                I   19
      IRET = IRET-(FLOW(M)*COST(M))*10000                               I   20
 20   CONTINUE                                                          I   21
      RETURN                                                            I   23
      END                                                               I   26-
```

```
                    SUBROUTINE NETFLO
C                   ************************                    J   2
C**** DIMENSION , COMMON , INTEGER , REAL AND LOGICAL STATEMENTS       J   3
      COMMON /UNO/ I(250), J(250), COST(250), LO(250), FLOW(250), PI(250  J   4
     1)                                                                J   5
      COMMON /UNOSA/ HI(300)                                           J   6
      COMMON /DOS/ NODES, ARCS, INFES, IRET                            J   7
      DIMENSION NA(300), NB(300)                                       J   8
      LOGICAL INFES                                                    J   9
      INTEGER A, AOK, C, COK, DEL, E, EPS, INF, LAB, N, NI, NJ, SRC, SNK  J  10
     1, FLOW, PI, NA, NODES, ARCS, I, J, COST, HI, LO, NB              J  11
C**** CHECK FEASIBILITY OF FORMULATION                                 J  12
      INFES = .TRUE.                                                   J  13
      DO 10 A = 1, ARCS                                                J  14
      IF (LO(A) .GT. HI(A)) GO TO 200                                  J  15
 10   CONTINUE                                                         J  16
C**** SET INF TO MAX AVAILABLE INTEGER                                 J  17
      INF = 999999                                                     J  18
      AOK = 0                                                          J  19
C**** FIND OUT OF KILTER ARC                                           J  20
 20   DO 30 A = 1, ARCS                                                J  21
      IA = I(A)$JA = J(A)                                              J  22
      C = COST(A)+PI(IA)-PI(JA)                                        J  23
C**** ARC IS IN STATE A1,B1 OR C1
      IF ((FLOW(A) .LT. LO(A)) .OR. (C .LT. 0 .AND. FLOW(A) .LT. HI(A)))  J  24
     1 GO TO 40                                                        J  25
C**** ARC IS NI STATE A2,B2 OR C2
      IF ((FLOW(A) .GT. HI(A)) .OR. (C .GT. 0 .AND. FLOW(A) .GT. LO(A)))  J  26
     1 GO TO 50                                                        J  27
 30   CONTINUE                                                         J  28
C**** NO REMAINING OUT OF KILTER ARCS                                  J  29
      GO TO 230                                                        J  30
C**** FLOW MUST BE INCREASED IN THE ARC TO BRING IT INTO KILTER
 40   SRC = J(A)                                                       J  31
      SNK = I(A)                                                       J  32
      E = +1                                                           J  33
      GO TO 60                                                         J  34
C**** FLOW MUST BE DECREASED IN THE ARC TO BRING IT INTO KILTER
 50   SRC = I(A)                                                       J  35
      SNK = J(A)                                                       J  36
      E = -1                                                           J  37
      GO TO 60                                                         J  38
C**** ATTEMPT TO BRING OUT OF KILTER ARCS INTO KILTER                  J  39
 60   IF ((A .EQ. AOK) .AND. (NA(SRC) .NE. 0)) GO TO 80                J  40
      AOK = A                                                          J  41
      DO 70 N = 1, NODES                                               J  42
      NA(N) = 0                                                        J  43
 70   NB(N) = 0                                                        J  44
      NA(SRC) = IABS(SNK)*E                                            J  45
      NB(SRC) = IABS(AOK)*E                                            J  46
 80   COK = C                                                          J  47
 90   LAB = 0                                                          J  48
      DO 120 A = 1, ARCS                                               J  49
      IA = I(A)$JA = J(A)                                              J  50
C**** BOTH NODES OF THE ARC ARE EITHER LABFLED OR UNLABELED
      IF ((NA(IA) .EQ. 0 .AND. NA(JA) .EQ. 0) .OR. (NA(IA) .NE. 0 .. AND  J  51
     1.NA(JA) .NE. 0)) GO TO 120                                       J  52
      C = COST(A)+PI(IA)-PI(JA)                                        J  53
C**** I-TH NODE IS NOT LABELED
      IF (NA(IA) .EQ. 0) GO TO 100                                     J  54
C**** ARC FLOW CANNOT BE INCREASED WITHOUT DRIVING IT (MORE) OUT OF KILTER
```

```
      IF (FLOW(A) .GE. HI(A) .OR. (FLOW(A) .GE. LO(A) .AND. C .GT. 0)) G  J  55
     1O TO 120                                                            J  56
C**** LABELING PROCESS THROUGH A FORWARD ARC
      NA(JA) = I(A)                                                       J  57
      NB(JA) = A                                                          J  58
      GO TO 110                                                           J  59
C**** ARC FLOW CANNOT BE DECREASED WITHOUT DRIVING IT (MORE) OUT OF KILTER
 100  IF (FLOW(A) .LE. LO(A) .OR. (FLOW(A) .LE. HI(A) .AND. C .LT. 0)) G  J  60
     1O TO 120                                                            J  61
C**** LABELING PROCESS THROUGH A REVERSE ARC
      IA = I(A)                                                           J  62
      NA(IA) = -J(A)                                                      J  63
      NB(IA) = -A                                                         J  64
 110  LAB = 1                                                             J  65
C**** NODE LABELED, TEST FOR BREAKTHRU                                     J  66
      IF (NA(SNK) .NE. 0) GO TO 150                                       J  67
 120  CONTINUE                                                            J  68
C**** NO BREAKTHRU                                                        J  69
      IF (LAB .NE. 0) GO TO 90                                            J  70
C**** DETERMINE CHANGE TO PI VECTOR                                       J  71
      DEL = INF                                                           J  72
      DO 130 A = 1, ARCS                                                  J  73
      IA = I(A)$JA = J(A)                                                 J  74
      IF ((NA(IA) .EQ. 0 .AND. NA(JA) .EQ. 0) .OR. (NA(IA) .NE. 0 .AND.   J  75
     1NA(JA) .NE. 0)) GO TO 130                                           J  76
      C = COST(A)+PI(IA)-PI(JA)                                           J  77
      IF (NA(JA) .EQ. 0 .AND. FLOW(A) .LT. HI(A)) DEL = MINO(DEL, C)      J  78
      IF (NA(JA) .NE. 0 .AND. FLOW(A) .GT. LO(A)) DEL = MINO(DEL, -C)     J  79
 130  CONTINUE                                                            J  80
      IF (DEL .EQ. INF .AND. (FLOW(AOK) .EQ. HI(AOK) .OR. FLOW(AOK) .EQ.  J  81
     1 LO(AOK))) DEL = ABS(COK)                                           J  82
C**** EXIT, NO FEASIBLE FLOW PATTERN                                      J  83
      IF (DEL .EQ. INF) GO TO 210                                         J  84
C**** CHANGE PI VECTOR BY COMPUTED DEL                                    J  85
      DO 140 N = 1, NODES                                                 J  86
 140  IF (NA(N) .EQ. 0) PI(N) = PI(N)+DEL                                 J  87
C**** FIND ANOTHER OUT-OF-KILTER ARC                                      J  88
      GO TO 20                                                            J  89
C**** BREAKTHRU, COMPUTE INCREMENTAL FLOW                                 J  90
 150  EPS = INF                                                           J  91
      NI = SRC                                                            J  92
 160  NJ = IABS(NA(NI))                                                   J  93
      A = IABS(NB(NI))                                                    J  94
      C = COST(A)+PI(NI)-PI(NJ)                                           J  95
      IF (NB(NI) .LT. 0) GO TO 170                                        J  96
      IF (C .GT. 0 .AND. FLOW(A) .LT. LO(A)) EPS = MINO(EPS, LO(A)-FLOW(  J  97
     1A))                                                                 J  98
      IF (C .LE. 0 .AND. FLOW(A) .LT. HI(A)) EPS = MINO(EPS, HI(A)-FLOW(  J  99
     1A))                                                                 J 100
      GO TO 180                                                           J 101
 170  IF (C .LT. 0 .AND. FLOW(A) .GT. HI(A)) EPS = MINO(EPS, FLOW(A)-HI(  J 102
     1A))                                                                 J 103
      IF (C .GE. 0 .AND. FLOW(A) .GT. LO(A)) EPS = MINO(EPS, FLOW(A)-LO(  J 104
     1A))                                                                 J 105
 180  NI = NJ                                                             J 106
      IF (NI .NE. SRC) GO TO 160                                          J 107
C**** CHANGE FLOW VECTOR BY COMPUTED EPS                                  J 108
 190  NJ = IABS(NA(NI))                                                   J 109
      A = IABS(NB(NI))                                                    J 110
      FLOW(A) = FLOW(A)+ISIGN(EPS, NB(NI))                                J 111
      NI = NJ                                                             J 112
      IF (NI .NE. SRC) GO TO 190                                          J 113
```

```
C**** FIND ANOTHER OUT OF KILTER ARC                                     J 114
      AOK = 0                                                             J 115
      GO TO 20                                                            J 116
  200 PRINT 240, A                                                        J 117
      PRINT 260, ARCS                                                     J 118
      GO TO 220                                                           J 119
  210 PRINT 250, AOK                                                      J 120
  220 INFES = .FALSE.                                                     J 121
  230 CONTINUE                                                            J 122
      RETURN                                                              J 123
C                                                                         J 124
  240 FORMAT (//,*   LOWER BOUND IS HIGHER THAN UPPER BOUND FOR ARC *,I5  J 125
     1)                                                                   J 126
  250 FORMAT (//,*   ARC*,I5,* IS OUT OF KILTER*)                          J 127
  260 FORMAT (//,* NUMBER OF ARCS(ACCORDING TO NETFLO ) =*I5)             J 128
      END                                                                 J 129-


                        SUBROUTINE EXTDAT
C                       ************************                          F   2
C**** DIMENSION , COMMON , INTEGER , REAL AND LOGICAL STATEMENTS           F   3
      COMMON /UNO/ I(250), J(250), COST(250), LO(250), FLOW(250), PI(250  F   4
     1)                                                                   F   5
      COMMON /UNOS/ AOPCST(20)                                            F   6
      COMMON /UNOSA/ HI(300)                                              F   7
      COMMON /DOS/ NODES, ARCS, INFES, IRET                               F   8
      COMMON /TRES/ DEBT, DELIM, LL, MM, N, PMAX, R, R1, SS, SUMA, T, TM  F   9
     1AX, TOT                                                             F  10
      COMMON /CUATRO/ AVCAP(50), CAPAC(20), INV(20), LIFE(20), MAR(20),   F  11
     1NONO(20), NRES(20), PVEARN(50)                                      F  12
      COMMON /CINCO/ OPTIME                                               F  13
      COMMON /SEIS/ BND(20, 50), IGORET(50), PERBAR(20, 50), PVBAR(20, 5  F  14
     10)                                                                  F  15
      INTEGER SS                                                          F  16
      INTEGER T, TMAX, TOT                                                F  17
      LOGICAL OPTIME                                                      F  18
C**** INITIAL CONDITIONS                                                   F  19
      OPTIME = .TRUE.                                                     F  20
      IGORET(1) = 0                                                       F  21
      MM = 14                                                             F  22
      N = 5                                                               F  23
      R = 0.04625                                                         F  24
      R1 = 0.1                                                            F  25
      SS = MM+1                                                           F  26
      T = 1                                                               F  27
      TMAX = 50                                                           F  28
      CUMDIV = 0.                                                         F  29
      LL = N+1                                                            F  30
      SUMA = 0                                                            F  31
C**** DATA ON EXISTING RESERVOIRS                                          F  32
      READ 10, (NRES(K), NONO(K), CAPAC(K), AOPCST(K), K = 1, N)          F  33
C**** DATA ON POTENTIAL NEW RESERVOIRS                                     F  34
      READ 20, (NRES(K), NONO(K), CAPAC(K), INV(K), MAR(K), AOPCST(K), L  F  35
     1IFE(K), K = LL, MM)                                                 F  36
C**** FINANCIAL DATA                                                       F  37
      READ 30, AVCAP(1)                                                   F  38
C**** FORMAT STATEMENTS                                                    F  39
      RETURN                                                              F  40
C                                                                         F  41
   10 FORMAT   (2I5,2F10.0)                                               F  42
   20 FORMAT   (2I5,F10.0,2I10,F10.0,I5)                                  F  43
   30 FORMAT   (F12.2)                                                    F  44
      END                                                                 F  45-
```

Appendix B

DETAILED SOLUTION OF THE EXAMPLE PROBLEM

B.1. Initial Data

Table B.1

Optimization of the Expansion of an Existing Water Resources System

EXISTING RESERVOIRS
****

RESERVOIR NUMBER	REPRESENTED BY NODE	CAPACITY (ACRE-FT.)	ANNUAL OPERATING COSTS
1	1	40	-0
2	4	30	-0
3	16	18	-0
4	22	30	-0
5	28	40	-0

NEW RESERVOIRS
****

RESERVOIR NUMBER	REPRESENTED BY NODE	CAPACITY (ACRE-FT.)	REQUIRED INVESTMENT	ANNUAL OPERATING COSTS	MAX.ANNUAL RETURN	USEFUL LIFE(YRS.)
6	14	60	40000000	-0	2700000	50
7	14	50	38000000	-0	2430000	50
8	14	40	34000000	-0	2160000	50
9	21	12	6300000	-0	360000	50
10	21	10	5500000	-0	300000	50
11	21	8	4550000	-0	240000	50
12	27	20	10000000	-0	600000	50
13	27	15	8400000	-0	480000	50
14	27	10	5550000	-0	300000	50

Table B.2

The PVBAR Matrix[a]

11747025	8660733	7486031	642335	296331	93695	1548456	856447	248541
11227742	8277881	7155107	613941	283231	89554	1480006	818587	237554
10731414	7911953	6838812	586801	270711	85595	1414581	782401	227053
10257026	7562202	6536499	560861	258744	81811	1352049	747815	217016
9803609	7227911	6247550	536068	247306	78195	1292281	714757	207423
9370236	6908397	5971374	512371	236374	74738	1235155	683161	198253
8956020	6603008	5707406	489721	225925	71434	1180555	652962	189490
8560115	6311119	5455107	468073	215938	68276	1128368	624097	181113
8181711	6032133	5213962	447381	206392	65258	1078487	596509	173107
7820034	5765479	4983476	427605	197268	62373	1030812	570140	165455
7474346	5510614	4763179	408702	188548	59616	985245	544936	158141
7143938	5267014	4552620	390635	180213	56981	941692	520847	151150
6828137	5034183	4351369	373367	172247	54462	900064	497823	144468
6526296	4811645	4159015	356862	164633	52054	860276	475816	138082
6237798	4598943	3975164	341087	157355	49753	822247	454783	131978
5962053	4395645	3799440	326009	150399	47554	785899	434679	126144
5698497	4201333	3631483	311598	143750	45452	751158	415464	120568
5446592	4015611	3470952	297823	137396	43443	717953	397098	115238
5205823	3838099	3317517	284658	131322	41522	686215	379544	110144
4975697	3668434	3170864	272074	125517	39687	655881	362766	105275
4755744	3506269	3030695	260047	119969	37932	626887	346730	100621
4545514	3351273	2896721	248552	114665	36255	599175	331402	96173
4344577	3203128	2768670	237564	109596	34653	572689	316753	91922
4152523	3061532	2646280	227063	104752	33121	547373	302750	87858
3968959	2926196	2529300	217025	100121	31657	523176	289367	83974
3793509	2796842	2417491	207432	95695	30257	500048	276576	80262
3625815	2673206	2310624	198262	91465	28920	477944	264349	76714
3465534	2555036	2208482	189498	87422	27641	456816	252664	73323
3312338	2442089	2110855	181121	83557	26420	436622	241495	70082
3165915	2334135	2017544	173114	79863	25252	417321	230819	66984
3025964	2230954	1928357	165462	76333	24135	398873	220616	64023
2892200	2132333	1843113	158147	72959	23068	381241	210863	61193
2764349	2038072	1761637	151156	69734	22049	364388	201542	58487
2642149	1947978	1683763	144475	66651	21074	348280	192633	55902
2525352	1861867	1609332	138088	63705	20142	332884	184117	53431
2413717	1779562	1538190	131984	60888	19252	318169	175978	51069
2307018	1700896	1470194	126149	58197	18401	304104	168199	48811
2205035	1625707	1405203	120573	55624	17588	290661	160764	46654
2107560	1553842	1343086	115243	53165	16810	277812	153657	44591
2014394	1485153	1283714	110148	50815	16067	265531	146865	42620
1925347	1419501	1226967	105279	48569	15357	253793	140372	40736
1840236	1356752	1172728	100625	46422	14678	242574	134167	38935
1758888	1296776	1120887	96177	44370	14029	231851	128236	37214
1681135	1239451	1071338	91926	42408	13409	221602	122567	35569
1606820	1184661	1023979	87862	40534	12816	211806	117149	33997
1535789	1132292	978713	83978	38742	12250	202443	111971	32494
1467899	1082239	935449	80266	37029	11708	193494	107021	31057
1403010	1034398	894097	76718	35392	11191	184940	102290	29685
1340989	988672	854573	73326	33828	10696	176765	97768	28372
1281710	944967	816796	70085	32332	10223	168951	93446	27118

[a] Columns correspond to new reservoirs (6–14), and rows correspond to years (1–50).

Table B.3

The Initial Network Configuration[a]

M	I	J	COST	HI	LO
1	239	1	-0	40	40
2	239	2	-0	15	15
3	239	4	-0	30	30
4	239	10	-0	10	10
5	239	14	-0	70	70
6	239	16	-0	12	12
7	239	21	-0	11	11
8	239	22	-0	20	20
9	239	27	-0	15	15
10	239	28	-0	30	30
11	239	1	-0	20	20
12	239	4	-0	15	15
13	239	14	-0	-0	-0
14	239	16	-0	9	9
15	239	21	-0	-0	-0
16	239	22	-0	30	30
17	239	27	-0	-0	-0
18	239	28	-0	40	40
19	1	2	-0	80	-0
20	2	3	-0	90	-0
21	3	10	-0	70	-0
22	4	5	-0	60	-0
23	5	6	-0	40	-0
24	6	8	-0	70	-0
25	8	10	-0	60	-0
26	10	11	-0	190	-0
27	11	12	-0	190	-0
28	14	15	-0	80	-0
29	15	12	-0	80	-0
30	12	18	-0	290	-0
31	16	18	-0	25	-0
32	18	19	-0	300	-0
33	19	26	-0	300	-0
34	22	24	-0	70	-0
35	21	24	-0	50	-0
36	24	26	-0	80	8
37	26	32	-0	330	-0
38	28	30	-0	60	-0
39	27	30	-0	50	-0
40	30	32	-0	95	8
41	5	7	-0	40	-0
42	7	6	-3	15	-0
43	3	9	-0	55	-0
44	8	9	-0	40	-0
45	9	10	-2	20	-0
46	11	13	-0	250	-0

[a] Number of nodes is 240, and number of arcs is 131.

Table B.3 ***(continued)***

M	I	J	COST	HI	LO
47	13	12	-4	-0	-0
48	15	13	-0	60	-0
49	16	17	-3	20	-0
50	19	20	-2	40	-0
51	22	23	-4	30	-0
52	24	25	-3	15	-0
53	28	29	-4	30	-0
54	30	31	-3	35	-0
55	1	41	-0	20	20
56	4	44	-0	15	15
57	14	54	-0	-0	-0
58	16	56	-0	18	-0
59	21	61	-0	-0	-0
60	22	62	-0	30	8
61	27	67	-0	-0	-0
62	28	68	-0	40	12
63	239	41	-0	60	60
64	239	42	-0	20	20
65	239	44	-0	50	50
66	239	50	-0	30	30
67	239	54	-0	60	60
68	239	56	-0	25	25
69	239	61	-0	15	15
70	239	62	-0	40	40
71	239	67	-0	30	30
72	239	68	-0	60	60
73	41	42	-0	80	-0
74	42	43	-0	90	-0
75	43	50	-0	70	-0
76	44	45	-0	80	-0
77	45	46	-0	60	-0
78	46	48	-0	70	-0
79	48	50	-0	60	-0
80	50	51	-0	190	-0
81	51	52	-0	190	-0
82	54	55	-0	60	-0
83	55	52	-0	60	-0
84	52	58	-0	290	-0
85	56	58	-0	25	-0
86	58	59	-0	300	-0
87	59	66	-0	300	-0
88	62	64	-0	70	-0
89	61	64	-0	50	-0
90	64	66	-0	80	8
91	66	72	-0	330	-0
92	68	70	-0	60	-0

Table B.3 (*continued*)

M	I	J	COST	HI	LO
93	67	70	-0	50	-0
94	70	72	-0	95	8
95	45	47	-0	40	-0
96	47	46	-4	15	-0
97	43	49	-0	55	-0
98	48	49	-0	40	-0
99	49	50	-3	20	-0
100	51	53	-0.	250	-0
101	53	52	-5	-0	-0
102	55	53	-0	60	-0
103	56	57	-0	-0	-0
104	59	60	-0	-0	-0
105	62	63	-0	-0	-0
106	64	65	-0	-0	-0
107	68	69	-0	-0	-0
108	70	71	-0	-0	-0
109	41	240	-0	20	20
110	44	240	-0	15	15
111	54	240	-0	-0	-0
112	56	240	-0	9	9
113	61	240	-0	-0	-0
114	62	240	-0	30	30
115	67	240	-0	-0	-0
116	68	240	-0	40	40
117	32	240	-0	400	-0
118	72	240	-0	400	-0
119	17	240	-0	20	-0
120	20	240	-0	40	-0
121	23	240	-0	30	-0
122	25	240	-0	90000	-0
123	29	240	-0	50	-0
124	31	240	-0	90000	-0
125	57	240	-0	20	-0
126	60	240	-0	40	-0
127	63	240	-0	30	-0
128	65	240	-0	15	-0
129	69	240	-0	50	-0
130	71	240	-0	35	-0
131	240	239	-0	90000	-0

Table B.4

The Operating Policy for the Original Network Configuration[a]

M	I	J	COST	HI	LO	FLOW
1	239	1	-0	40	40	40
2	239	2	-0	15	15	15
3	239	4	-0	30	30	30
4	239	10	-0	10	10	10
5	239	14	-0	70	70	70
6	239	16	-0	12	12	12
7	239	21	-0	11	11	11
8	239	22	-0	20	20	20
9	239	27	-0	15	15	15
10	239	28	-0	30	30	30
11	239	1	-0	20	20	20
12	239	4	-0	15	15	15
13	239	14	-0	-0	-0	-0
14	239	16	-0	9	9	9
15	239	21	-0	-0	-0	-0
16	239	22	-0	30	30	30
17	239	27	-0	-0	-0	-0
18	239	28	-0	40	40	40
19	1	2	-0	80	-0	40
20	2	3	-0	90	-0	55
21	3	10	-0	70	-0	50
22	4	5	-0	60	-0	30
23	5	6	-0	40	-0	15
24	6	8	-0	70	-0	30
25	8	10	-0	60	-0	15
26	10	11	-0	190	-0	95
27	11	12	-0	190	-0	95
28	14	15	-0	80	-0	70
29	15	12	-0	80	-0	70
30	12	18	-0	290	-0	165
31	16	18	-0	25	-0	1
32	18	19	-0	300	-0	166
33	19	26	-0	300	-0	126
34	22	24	-0	70	-0	12
35	21	24	-0	50	-0	11
36	24	26	-0	80	8	8
37	26	32	-0	330	-0	134
38	28	30	-0	60	-0	28
39	27	30	-0	50	-0	15
40	30	32	-0	95	8	8
41	5	7	-0	40	-0	15
42	7	6	-3	15	-0	15
43	3	9	-0	55	-0	5
44	8	9	-0	40	-0	15
45	9	10	-2	20	-0	20
46	11	13	-0	250	-0	-0

[a] Number of nodes is 240, and number of arcs is 131.

Table B.4 (*continued*)

M	I	J	COST	HI	LO	FLOW
47	13	12	-4	-0	-0	-0
48	15	13	-0	60	-0	-0
49	16	17	-3	20	-0	20
50	19	20	-2	40	-0	40
51	22	23	-4	30	-0	30
52	24	25	-3	15	-0	15
53	28	29	-4	30	-0	30
54	30	31	-3	35	-0	35
55	1	41	-0	20	20	20
56	4	44	-0	15	15	15
57	14	54	-0	-0	-0	-0
58	16	56	-0	18	-0	-0
59	21	61	-0	-0	-0	-0
60	22	62	-0	30	8	8
61	27	67	-0	-0	-0	-0
62	28	68	-0	40	12	12
63	239	41	-0	60	60	60
64	239	42	-0	20	20	20
65	239	44	-0	50	50	50
66	239	50	-0	30	30	30
67	239	54	-0	60	60	60
68	239	56	-0	25	25	25
69	239	61	-0	15	15	15
70	239	62	-0	40	40	40
71	239	67	-0	30	30	30
72	239	68	-0	60	60	60
73	41	42	-0	80	-0	60
74	42	43	-0	90	-0	80
75	43	50	-0	70	-0	70
76	44	45	-0	80	-0	50
77	45	46	-0	60	-0	35
78	46	48	-0	70	-0	50
79	48	50	-0	60	-0	40
80	50	51	-0	190	-0	160
81	51	52	-0	190	-0	160
82	54	55	-0	60	-0	60
83	55	52	-0	60	-0	60
84	52	58	-0	290	-0	220
85	56	58	-0	25	-0	16
86	58	59	-0	300	-0	236
87	59	66	-0	300	-0	236
88	62	64	-0	70	-0	18
89	61	64	-0	50	-0	15
90	64	66	-0	80	8	33
91	66	72	-0	330	-0	269
92	68	70	-0	60	-0	32

Table B.4 (*continued*)

M	I	J	COST	HI	LO	FLOW
93	67	70	-0	50	-0	30
94	70	72	-0	95	8	62
95	45	47	-0	40	-0	15
96	47	46	-4	15	-0	15
97	43	49	-0	55	-0	10
98	48	49	-0	40	-0	10
99	49	50	-3	20	-0	20
100	51	53	-0	250	-0	-0
101	53	52	-5	-0	-0	-0
102	55	53	-0	60	-0	-0
103	56	57	-0	-0	-0	-0
104	59	60	-0	-0	-0	-0
105	62	63	-0	-0	-0	-0
106	64	65	-0	-0	-0	-0
107	68	69	-0	-0	-0	-0
108	70	71	-0	-0	-0	-0
109	41	240	-0	20	20	20
110	44	240	-0	15	15	15
111	54	240	-0	-0	-0	-0
112	56	240	-0	9	9	9
113	61	240	-0	-0	-0	-0
114	62	240	-0	30	30	30
115	67	240	-0	-0	-0	-0
116	68	240	-0	40	40	40
117	32	240	-0	400	-0	142
118	72	240	-0	400	-0	331
119	17	240	-0	20	-0	20
120	20	240	-0	40	-0	40
121	23	240	-0	30	-0	30
122	25	240	-0	90000	-0	15
123	29	240	-0	50	-0	30
124	31	240	-0	90000	-0	35
125	57	240	-0	20	-0	-0
126	60	240	-0	40	-0	-0
127	63	240	-0	30	-0	-0
128	65	240	-0	15	-0	-0
129	69	240	-0	50	-0	-0
130	71	240	-0	35	-0	-0
131	240	239	-0	90000	-0	757

Table B.4 (*continued*)

M	1	2	3	4	5	6	7	8	9	10	11	12	13	14
PYE(M)	0	0	0	0	0	0	0	0	0	0	0	0	0	0
M	15	16	17	18	19	20	21	22	23	24	25	26	27	28
PYE(M)	0	0	0	0	0	0	0	0	0	0	0	0	0	0
M	29	30	31	32	33	34	35	36	37	38	39	40	41	42
PYE(M)	0	0	0	0	0	0	0	0	0	0	0	0	0	0
M	43	44	45	46	47	48	49	50	51	52	53	54	55	56
PYE(M)	0	0	0	0	0	0	0	0	0	0	0	0	0	0
M	57	58	59	60	61	62	63	64	65	66	67	68	69	70
PYE(M)	0	0	0	0	0	0	0	0	0	0	0	0	0	0
M	71	72	73	74	75	76	77	78	79	80	81	82	83	84
PYE(M)	0	0	0	0	0	0	0	0	0	0	0	0	0	0
M	85	86	87	88	89	90	91	92	93	94	95	96	97	98
PYE(M)	0	0	0	0	0	0	0	0	0	0	0	0	0	0
M	99	100	101	102	103	104	105	106	107	108	109	110	111	112
PYE(M)	0	0	0	0	0	0	0	0	0	0	0	0	0	0
M	113	114	115	116	117	118	119	120	121	122	123	124	125	126
PYE(M)	0	0	0	0	0	0	0	0	0	0	0	0	0	0
M	127	128	129	130	131									
PYE(M)	0	0	0	0	0									

B.2. Intermediate Results

Table B.5

Feasible Solution

```
   FEASIBLE SOLUTION NUMBER    1
***********************************


DAM  9 WAS CONSTRUCTED AND BEGAN OPERATION IN YEAR  2
DAM 12 WAS CONSTRUCTED AND BEGAN OPERATION IN YEAR  3
DAM  8 WAS CONSTRUCTED AND BEGAN OPERATION IN YEAR  9

THE TOTAL PRESENT VALUE OF EARNINGS = 147529961DOLLARS
```

Table B.6

The Bound Matrix[a]

0	0	0	0	0	0	0	0	0	0
0	152599208	151476433	0	154712578	154518900	0	154456746	153875713	154965415
0	152208459	151135317	0	0	0	0	153983884	153428536	154493764
0	0	0	0	0	0	0	0	0	0
0	0	0	0	0	0	0	0	0	0
0	0	0	0	0	0	0	0	0	0
0	0	0	0	0	0	0	0	0	0
0	0	0	0	0	0	0	0	0	0
0	0	0	0	0	0	0	0	0	150444348
0	0	0	0	0	0	0	0	0	0
0	0	0	0	0	0	0	0	0	0
0	0	0	0	0	0	0	0	0	0
0	0	0	0	0	0	0	0	0	0
0	0	0	0	0	0	0	0	0	0
0	0	0	0	0	0	0	0	0	0
0	0	0	0	0	0	0	0	0	0
0	0	0	0	0	0	0	0	0	0
0	0	0	0	0	0	0	0	0	0
0	0	0	0	0	0	0	0	0	0
0	0	0	0	0	0	0	0	0	0
0	0	0	0	0	0	0	0	0	0
0	0	0	0	0	0	0	0	0	0
0	0	0	0	0	0	0	0	0	0
0	0	0	0	0	0	0	0	0	0
0	0	0	0	0	0	0	0	0	0

0	0	0	0	0	0	0	0	0	0
0	0	0	0	0	0	0	0	0	0
0	0	0	0	0	0	0	0	0	0
0	0	0	0	0	0	0	0	0	0
0	0	0	0	0	0	0	0	0	0
0	0	0	0	0	0	0	0	0	0
0	0	0	0	0	0	0	0	0	0
0	0	0	0	0	0	0	0	0	0
0	0	0	0	0	0	0	0	0	0
0	0	0	0	0	0	0	0	0	0
0	0	0	0	0	0	0	0	0	0
0	0	0	0	0	0	0	0	0	0
0	0	0	0	0	0	0	0	0	0
0	0	0	0	0	0	0	0	0	0
0	0	0	0	0	0	0	0	0	0
0	0	0	0	0	0	0	0	0	0
0	0	0	0	0	0	0	0	0	0
0	0	0	0	0	0	0	0	0	0
0	0	0	0	0	0	0	0	0	0
0	0	0	0	0	0	0	0	0	0
0	0	0	0	0	0	0	0	0	0
0	0	0	0	0	0	0	0	0	0
0	0	0	0	0	0	0	0	0	0
0	0	0	0	0	0	0	0	0	0
0	0	0	0	0	0	0	0	0	0

[a] Columns 1–9 correspond to new reservoirs (6–14), column 10 corresponds to no new reservoir, and rows correspond to years (1–50).

Table B.7

Feasible Solutions

```
   FEASIBLE SOLUTION NUMBER    2
**********************************


DAM  9 WAS CONSTRUCTED AND BEGAN OPERATION IN YEAR  2
DAM 12 WAS CONSTRUCTED AND BEGAN OPERATION IN YEAR  3
DAM  6 WAS CONSTRUCTED AND BEGAN OPERATION IN YEAR 10

THE TOTAL PRESENT VALUE OF EARNINGS = 149829858DOLLARS


   FEASIBLE SOLUTION NUMBER    3
**********************************


DAM  9 WAS CONSTRUCTED AND BEGAN OPERATION IN YEAR  2
DAM 12 WAS CONSTRUCTED AND BEGAN OPERATION IN YEAR  4
DAM  6 WAS CONSTRUCTED AND BEGAN OPERATION IN YEAR 10

THE TOTAL PRESENT VALUE OF EARNINGS = 150111063DOLLARS


   FEASIBLE SOLUTION NUMBER    4
**********************************


DAM  9 WAS CONSTRUCTED AND BEGAN OPERATION IN YEAR  2
DAM 12 WAS CONSTRUCTED AND BEGAN OPERATION IN YEAR  5
DAM  6 WAS CONSTRUCTED AND BEGAN OPERATION IN YEAR 10

THE TOTAL PRESENT VALUE OF EARNINGS = 150329764DOLLARS


   FEASIBLE SOLUTION NUMBER    5
**********************************


DAM  9 WAS CONSTRUCTED AND BEGAN OPERATION IN YEAR  2
DAM 12 WAS CONSTRUCTED AND BEGAN OPERATION IN YEAR  6
DAM  6 WAS CONSTRUCTED AND BEGAN OPERATION IN YEAR 10

THE TOTAL PRESENT VALUE OF EARNINGS = 150490937DOLLARS
```

Table B.7 ***(continued)***

```
   FEASIBLE SOLUTION NUMBER    6
*****************************************


DAM   9 WAS CONSTRUCTED AND BEGAN OPERATION IN YEAR  2
DAM  12 WAS CONSTRUCTED AND BEGAN OPERATION IN YEAR  7
DAM   6 WAS CONSTRUCTED AND BEGAN OPERATION IN YEAR 10


THE TOTAL PRESENT VALUE OF EARNINGS = 150599241DOLLARS



   FEASIBLE SOLUTION NUMBER    7
*****************************************


DAM   9 WAS CONSTRUCTED AND BEGAN OPERATION IN YEAR  2
DAM   6 WAS CONSTRUCTED AND BEGAN OPERATION IN YEAR  8
DAM  12 WAS CONSTRUCTED AND BEGAN OPERATION IN YEAR 10


THE TOTAL PRESENT VALUE OF EARNINGS = 151918402DOLLARS



   FEASIBLE SOLUTION NUMBER    8
*****************************************


DAM   9 WAS CONSTRUCTED AND BEGAN OPERATION IN YEAR  3
DAM   6 WAS CONSTRUCTED AND BEGAN OPERATION IN YEAR  8
DAM  12 WAS CONSTRUCTED AND BEGAN OPERATION IN YEAR 10


THE TOTAL PRESENT VALUE OF EARNINGS = 152118006DOLLARS



   FEASIBLE SOLUTION NUMBER    9
*****************************************


DAM   9 WAS CONSTRUCTED AND BEGAN OPERATION IN YEAR  4
DAM   6 WAS CONSTRUCTED AND BEGAN OPERATION IN YEAR  8
DAM  12 WAS CONSTRUCTED AND BEGAN OPERATION IN YEAR 10


THE TOTAL PRESENT VALUE OF EARNINGS = 152256397DOLLARS
```

Table B.7 *(continued)*

```
   FEASIBLE SOLUTION NUMRER  1n
***********************************


DAM  9 WAS CONSTRUCTED AND BEGAN OPERATION IN YEAR  5
DAM  6 WAS CONSTRUCTED AND BEGAN OPERATION IN YEAR  8
DAM 12 WAS CONSTRUCTED AND BEGAN OPERATION IN YEAR 10

THE TOTAL PRESENT VALUE OF EARNINGS = 152338597DOLLARS


   FEASIBLE SOLUTION NUMRER  11
***********************************


DAM  9 WAS CONSTRUCTED AND BEGAN OPERATION IN YEAR  6
DAM  6 WAS CONSTRUCTED AND BEGAN OPERATION IN YEAR  8
DAM 12 WAS CONSTRUCTED AND BEGAN OPERATION IN YEAR 10

THE TOTAL PRESENT VALUE OF EARNINGS = 152369303DOLLARS


   FEASIBLE SOLUTION NUMRER  12
***********************************


DAM  6 WAS CONSTRUCTED AND BEGAN OPERATION IN YEAR  7
DAM  9 WAS CONSTRUCTED AND BEGAN OPERATION IN YEAR  8
DAM 12 WAS CONSTRUCTED AND BEGAN OPERATION IN YEAR 10

THE TOTAL PRESENT VALUE OF EARNINGS = 152972507DOLLARS
```

B.3. Final Results

Table B.8

The Bound Matrix[a,b]

0	0	0	0	0	0	0	0	0	0
0	0	0	0	0	0	0	0	0	0
0	0	0	0	0	0	0	0	0	0
0	0	0	0	0	0	0	0	0	0
0	0	0	0	0	0	0	0	0	0
0	0	0	0	0	0	0	0	0	0
0	0	0	0	0	0	0	0	0	0
0	0	0	0	0	0	0	0	0	0
0	0	0	0	0	0	0	0	0	0
0	0	0	0	0	0	0	0	0	0
0	0	0	0	0	0	0	0	0	0
0	0	0	0	0	0	0	0	0	0
0	0	0	0	0	0	0	0	0	0
0	0	0	0	0	0	0	0	0	0
0	0	0	0	0	0	0	0	0	0
0	0	0	0	0	0	0	0	0	0
0	0	0	0	0	0	0	0	0	0
0	0	0	0	0	0	0	0	0	0
0	0	0	0	0	0	0	0	0	0
0	0	0	0	0	0	0	0	0	0
0	0	0	0	0	0	0	0	0	0
0	0	0	0	0	0	0	0	0	0
0	0	0	0	0	0	0	0	0	0
0	0	0	0	0	0	0	0	0	0
0	0	0	0	0	0	0	0	0	0
0	0	0	0	0	0	0	0	0	0
0	0	0	0	0	0	0	0	0	0
0	0	0	0	0	0	0	0	0	0
0	0	0	0	0	0	0	0	0	0
0	0	0	0	0	0	0	0	0	0
0	0	0	0	0	0	0	0	0	0
0	0	0	0	0	0	0	0	0	0
0	0	0	0	0	0	0	0	0	0
0	0	0	0	0	0	0	0	0	0
0	0	0	0	0	0	0	0	0	0
0	0	0	0	0	0	0	0	0	0
0	0	0	0	0	0	0	0	0	0
0	0	0	0	0	0	0	0	0	0
0	0	0	0	0	0	0	0	0	0
0	0	0	0	0	0	0	0	0	0
0	0	0	0	0	0	0	0	0	0
0	0	0	0	0	0	0	0	0	0
0	0	0	0	0	0	0	0	0	0
0	0	0	0	0	0	0	0	0	0
0	0	0	0	0	0	0	0	0	0
0	0	0	0	0	0	0	0	0	0
0	0	0	0	0	0	0	0	0	0
0	0	0	0	0	0	0	0	0	0
0	0	0	0	0	0	0	0	0	0
0	0	0	0	0	0	0	0	0	0

[a] Columns 1–9 correspond to new reservoirs (6–14), column 10 corresponds to no new reservoir, and rows correspond to years (1–50).

[b] Since all elements are equal to zero, we thus have searched through every branch of the tree.

Table B.9

Construction Timing and Total Earnings

```
                    TIMING OF CONSTRUCTION OF DAMS
                    ****************************************


DAM  6 WAS CONSTRUCTED AND BEGAN OPERATION IN YEAR  7
DAM  9 WAS CONSTRUCTED AND BEGAN OPERATION IN YEAR  8
DAM 12 WAS CONSTRUCTED AND BEGAN OPERATION IN YEAR 10



                          TOTAL EARNINGS
                    *********************


THE TOTAL PRESENT VALUE OF NETT EARNINGS = 152972507DOLLARS


THE SYSTEM WAS OPERATED FOR 50YEARS
```

Table B.10

The Final Network Configuration[a]

M	I	J	COST	HI	LO	FLOW
1	239	1	-0	40	40	40
2	239	2	-0	15	15	15
3	239	4	-0	30	30	30
4	239	10	-0	10	10	10
5	239	14	-0	70	70	70
6	239	16	-0	12	12	12
7	239	21	-0	11	11	11
8	239	22	-0	20	20	20
9	239	27	-0	15	15	15
10	239	28	-0	30	30	30
11	239	1	-0	20	20	20
12	239	4	-0	15	15	15
13	239	14	-0	30	30	30
14	239	16	-0	9	9	9
15	239	21	-0	12	12	12
16	239	22	-0	30	30	30
17	239	27	-0	20	20	20
18	239	28	-0	40	40	40
19	1	2	-0	80	-0	40
20	2	3	-0	90	-0	55
21	3	10	-0	70	-0	50
22	4	5	-0	60	-0	30
23	5	6	-0	40	-0	15
24	6	8	-0	70	-0	30
25	8	10	-0	60	-0	15
26	10	11	-0	190	-0	95
27	11	12	-0	190	-0	65
28	14	15	-0	80	-0	70
29	15	12	-0	80	-0	70
30	12	18	-0	290	-0	165
31	16	18	-0	25	-0	1
32	18	19	-0	300	-0	166
33	19	26	-0	300	-0	126
34	22	24	-0	70	-0	12
35	21	24	-0	50	-0	23
36	24	26	-0	80	8	8
37	26	32	-0	330	-0	134
38	28	30	-0	60	-0	28
39	27	30	-0	50	-0	35
40	30	32	-0	95	8	8
41	5	7	-0	40	-0	15
42	7	6	-3	15	-0	15
43	3	9	-0	55	-0	5
44	8	9	-0	40	-0	15
45	9	10	-2	20	-0	20
46	11	13	-0	250	-0	30

[a] Number of rodes is 240, and number of arcs is 131.

Table B.10 (*continued*)

M	I	J	COST	HI	LO	FLOW
47	13	12	-4	30	-0	30
48	15	13	-0	60	-0	-0
49	16	17	-3	20	-0	20
50	19	20	-2	40	-0	40
51	22	23	-4	30	-0	30
52	24	25	-3	27	-0	27
53	28	29	-4	30	-0	30
54	30	31	-3	55	-0	55
55	1	41	-0	20	20	20
56	4	44	-0	15	15	15
57	14	54	-0	60	-0	30
58	16	56	-0	18	-0	-0
59	21	61	-0	12	-0	0
60	22	62	-0	30	8	8
61	27	67	-0	20	-0	0
62	28	68	-0	40	12	12
63	239	41	-0	60	60	60
64	239	42	-0	20	20	20
65	239	44	-0	50	50	50
66	239	50	-0	30	30	30
67	239	54	-0	60	60	60
68	239	56	-0	25	25	25
69	239	61	-0	15	15	15
70	239	62	-0	40	40	40
71	239	67	-0	30	30	30
72	239	68	-0	60	60	60
73	41	42	-0	80	-0	60
74	42	43	-0	90	-0	80
75	43	50	-0	70	-0	70
76	44	45	-0	80	-0	50
77	45	46	-0	60	-0	35
78	46	48	-0	70	-0	50
79	48	50	-0	60	-0	40
80	50	51	-0	190	-0	160
81	51	52	-0	190	-0	130
82	54	55	-0	60	-0	60
83	55	52	-0	60	-0	60
84	52	58	-0	290	-0	220
85	56	58	-0	25	-0	16
86	58	59	-0	300	-0	236
87	59	66	-0	300	-0	236
88	62	64	-0	70	-0	18
89	61	64	-0	50	-0	3
90	64	66	-0	80	8	21
91	66	72	-0	330	-0	257

Table B.10 ***(continued)***

M	I	J	COST	HI	LO	FLOW
92	68	70	-0	60	-0	32
93	67	70	-0	50	-0	10
94	70	72	-0	95	8	42
95	45	47	-0	40	-0	15
96	47	46	-4	15	-0	15
97	43	49	-0	55	-0	10
98	48	49	-0	40	-0	10
99	49	50	-3	20	-0	20
100	51	53	-0	250	-0	30
101	53	52	-5	30	-0	30
102	55	53	-0	60	-0	-0
103	56	57	-0	-0	-0	-0
104	59	60	-0	-0	-0	-0
105	62	63	-0	-0	-0	-0
106	64	65	-0	-0	-0	-0
107	68	69	-0	-0	-0	-0
108	70	71	-0	-0	-0	-0
109	41	240	-0	20	20	20
110	44	240	-0	15	15	15
111	54	240	-0	30	30	30
112	56	240	-0	9	9	9
113	61	240	-0	12	12	12
114	62	240	-0	30	30	30
115	67	240	-0	20	20	20
116	68	240	-0	40	40	40
117	32	240	-0	400	-0	142
118	72	240	-0	400	-0	299
119	17	240	-0	20	-0	20
120	20	240	-0	40	-0	40
121	23	240	-0	30	-0	30
122	25	240	-0	90000	-0	27
123	29	240	-0	50	-0	30
124	31	240	-0	90000	-0	55
125	57	240	-0	20	-0	-0
126	60	240	-0	40	-0	-0
127	63	240	-0	30	-0	-0
128	65	240	-0	15	-0	-0
129	69	240	-0	50	-0	-0
130	71	240	-0	35	-0	-0
131	240	239	-0	90000	-0	819

Table B.10 ***(continued)***

M	1	2	3	4	5	6	7	8	9	10	11	12	13	14
PYE(M)	0	0	0	0	0	0	0	0	0	0	0	0	0	0
M	15	16	17	18	19	20	21	22	23	24	25	26	27	28
PYE(M)	0	0	0	0	0	0	3	3	0	3	0	0	3	3
M	29	30	31	32	33	34	35	36	37	38	39	40	41	42
PYE(M)	0	3	0	0	6	6	6	6	6	6	6	6	0	0
M	43	44	45	46	47	48	49	50	51	52	53	54	55	56
PYE(M)	0	0	0	0	0	0	0	0	0	0	0	0	0	0
M	57	58	59	60	61	62	63	64	65	66	67	68	69	70
PYE(M)	6	0	0	6	0	0	6	0	6	0	0	0	6	0

Table B.10 ***(continued)***

M	71	72	73	74	75	76	77	78	79	80	81	82	83	84
PYE(M)	6	0	6	6	6	6	6	6	6	6	6	6	6	6

M	85	86	87	88	89	90	91	92	93	94	95	96	97	98
PYE(M)	6	6	6	6	6	6	6	6	6	6	6	6	6	6

M	99	100	101	102	103	104	105	106	107	108	109	110	111	112
PYE(M)	6	6	6	6	6	6	6	6	6	6	6	6	6	6

M	113	114	115	116	117	118	119	120	121	122	123	124	125	126
PYE(M)	6	6	6	6	6	6	6	6	6	6	6	6	6	6

M	127	128	129	130	131
PYE(M)	6	6	6	6	6

Appendix C

NOTATION

a	Constant	
$\mathbf{a}$	Vector of constants	
A	Subset of N in Section 3.3	
A_{jm}	$+1$ if flow in link (arc) m enters reservoir j, -1 if flow in link (arc) m leaves reservoir j, 0 otherwise	
A_m	Oxygen production ($A > 0$) or reduction ($A < 0$) due to plants and bottom deposits for each m	mg/l/day
b	Constant	
$\mathbf{b}$	Vector of constants	
b_{ij}	Benefit of passing one unit of flow through arc (i, j)	\$/acre-ft
B_{jt}	The bound associated with node j of year t	\$
BB_{imt}	BOD concentration at beginning of link m in month i of year t	mg/l
BBA	Branch and bound algorithm	
BE_{imt}	BOD concentration at end of link m in month i of year t	mg/l
BOD	Biological oxygen demand	
BT_{imt}	BOD concentration of tributary flow into link m in month i of year t	mg/l
BW_{ijt}	BOD concentration of waste water effluent discharged from plant j in month i of year t	mg/l
BW_{imt}	BOD concentration of total waste water effluents discharged into link m in month i of year t	mg/l
$BW_j^{\max}$	Maximum concentration of BOD discharged into link m from treatment facilities associated with plant j	mg/l
$BW_j^{\min}$	Minimum concentration of BOD discharged into link m from treatment facilities associated with plant j	mg/l
c	Constant	
c_{ij}	Cost of passing one unit of water flow between nodes i and j	\$/acre-ft
$\mathbf{c}$	Vector of constants	
c_{kl}	Capital cost per ton of crop k by method l	\$/ton
C_m	Capacity of arc m (a connecting link between reservoirs)	acre-ft

Symbol	Definition	Units
$\hat{C}_t$	Present value of available capital at beginning of year t	\$
C_{jt}	Capital needed for building project j (reservoir or treatment plant) in year t	\$
CB	Capital budgeting	
CB_{imt}	Dissolved oxygen concentration at beginning of link m in month i of year t	mg/l
CE_{imt}	Dissolved oxygen concentration at end of link m in month i of year t	mg/l
COD	Chemical oxygen demand	
CR_t	Cumulative net return from the system to the end of year t	\$
CS_{imt}	Saturation dissolved oxygen concentration in link m in month i of year t	mg/l
CT_{imt}	Dissolved oxygen concentration of tributary flow into link m in month i of year t	mg/l
CW_{imt}	Dissolved oxygen concentration of waste water effluent discharged into link m in month i of year t	mg/l
D_{ijt}	Amount of water supplied for irrigation from reservoir j in month i of year t	acre-ft
DB_{imt}	Dissolved oxygen deficit at beginning of link m in month i of year t	mg/l
DE_{imt}	Dissolved oxygen deficit at end of link m in month i of year t	mg/l
DO	Dissolved oxygen	
E_{ijt}	Losses from the water in storage in reservoir j during month i of year t	acre-ft
$f(\mathbf{x})$	Objective function	\$
f_{ij}	Flow in arc (i, j) in Section 3.3	acre-ft
F_m	Flow in arc m	acre-ft
$FE^{l}_{T\max}$	Net return from a feasible expansion policy over the time horizon	\$
$\tilde{F}_{it}$	Minimum municipal and industrial needs for a subsection of the river basin for month i of year t; $=$LO(M) for those arcs specifying municipal and industrial needs	acre-ft
F_{ijt}	Amount of water supplied for municipal and industrial use from reservoir j in month i of year t	acre-ft
$g_j(\mathbf{x})$	Inequality constraint	
$G\{N; A\}$	A collection of N nodes and a subset A of ordered pairs taken from N	
G_{it}	Minimum irrigation needs for a subsection of the river basin for month i of year t; $=$LO(M) for those arcs specifying irrigation needs	acre-ft
$h_j(\mathbf{x})$	Equality constraint	
H_{ijt}	Operating cost of supplying imported water to reservoir j in month i of year t	mills/acre-ft
$\hat{I}_t$	Investment in year t	\$
I_{ijt}	Amount of imported water supplied to reservoir j in month i of year t	acre-ft

J_{it}	Maximum available imported water for month i of year t	acre-ft
K_j	Amount of energy produced by turbine j per acre-ft of water	kW-hr/acre-ft
K_{jt}	Capital needed for building canal j in year t	$
$\tilde{K}_{ijt}$	Projected waste water flow for treatment plant j in month i of year t	acre-ft
l_{ij}	L_m in the OKA in Section 3.3	acre-ft
L_m	Lower bound on arc m	acre-ft
LBBA	Little's branch and bound algorithm	
M_t	Capital budget limit for year t	$
M	Maximum number of projects (dams)	
M^*	Maximum number of treatment plants	
M_1	Total number of links (arcs) in the system	
MR_j	Maximum annual return available from operating reservoir j	$
N	Number of nodes in OKA in Section 3.3	
N	Number of projects (dams) that exist at the beginning of the planning horizon ($t = 0$)	
N^*	Number of treatment plants that exist at the beginning of the planning horizon ($t = 0$)	
N_{OLER}	Number of calls on subroutine OLERSEN	
NLP	Nonlinear programming	
NZC	Nonzero cost	
OKA	Out-of-kilter algorithm	
OP	Operational policy	
OR_{jt}	Net operational return for reservoir j when introduced at the beginning of year t and operated at maximum efficiency for the years t through $T_{\max}$	$
p_{kl}	Market price per ton of crop k produced by method l	$/ton
P_{it}	Minimum acceptable energy demand for the total system of month i of year t	kW-hr
PF_m	Penalty function associated with arc m	$
q_{ij}	Total cost of transporting one unit of flow from node i to node j (Section 3.3) $= \pi_i - \pi_j - b_{ij}$	$/acre-ft
q_{kl}	Unit net consumptive use of irrigation water for crop k by method l	acre/acre-ft
Q_{imt}	Flow in link m during month i of year t	acre-ft
QT_{imt}	Tributary flow into link m in month i of year t	acre-ft
QW_{ijt}	Waste water flow discharged from treatment plant j in month i of year t	acre-ft
QW_{imt}	Waste water flow discharged into link m in month i of year t	acre-ft
r	Interest rate	
R_{it}	Minimum acceptable recreation need for a section of the river basin for month i of year t	
R_m	BOD addition due to runoff and scour in link m	mg/l/day
$\tilde{S}_{j0}$	Specified initial volume (initial conditions for reservoir j)	acre-ft

$\breve{S}_{jr}$	Minimum acceptable recreation level for reservoir j, $=$LO(M) for recreation in the reservoir arc	acre-ft
S_{ijt}	Storage volume of water in reservoir j at the beginning of month i of year t (also termed carry-over storage), $=$FLOW(M) for arc corresponding to the carry-over storage	acre-ft
t_{C}	Critical time at which the dissolved oxygen deficit is maximum for reach r	days
T_m	Time water takes to flow from the beginning to the end of link m	days
$T_{\max}$	Length of the planning horizon	yr
TR_t	Return from operating the system as it exists at the beginning of year t, for the years t through $T_{\max}$	$
$t_{i\mathrm{BND}}$	Running time for subroutine BOUND	sec
t_{OLER}	Running time for subroutine OLER	sec
u_{ij}	Upper flow capacity of arc (i, j) in Section 3.3	acre-ft
$\mathbf{u}$	Vector of shadow prices	
U_{ijt}	Net unregulated flow to reservoir j in month i of year t (deterministic)	acre-ft
V_j	Maximum capacity of reservoir j, $=$HI(M) for the reservoir arc	acre-ft
$\mathbf{x}$	Vector whose elements are variables	
$\hat{X}_t$	Return from operating the system in year t $\hat{X}_t = \sum_{i=1}^{12} \sum_{j=1}^{M} X_{ijt}$	$
X_{ijt}	Economic return from reservoir j in month i of year t, or the operating cost of treatment plant j in month i of year t	$
$\mathbf{y}$	Vector of decision variables	
y_{kl}	Production of crop k by method l, including new irrigation water	tons/acre
$Y_{i\mathrm{BND}}$	Years each project is under consideration	yr
Z_{ijt}	Head of water in reservoir j in month i of year t	ft
ZC	Zero cost	

Greek

α	Discount factor $1/(1 + r)^t$, where r is the discount rate and t is the number of years	
β_{jt}	1 indicates a return is available from (the arcs corresponding to) project j (treatment plant or reservoir) in year t; 0 indicates no return is available	
γ_{ij}	Marginal value of increasing u_{ij} by one unit (in Section 3.3)	$/acre-ft
γ	Fraction of water Q_{ijt} passing through turbines	
δ_{ij}	Marginal value of decreasing l_{ij} by one unit (in Section 3.3)	$/acre-ft
δ_{ijt}	Net revenue coefficient for the irrigation water supplied using a predetermined crop mix by reservoir j in month i of year t	$/acre-ft

θ_m	Constant	
λ_{ijt}	1 if water is imported to reservoir j in month i of year t; 0 otherwise	
λ_{jt}	1 indicates that capital must be provided for building project j (dam/treatment plant) in year t; 0 otherwise	
μ_m	Constant	
ν_m	Constant	
π_j	Price of one unit of water at node j (in Section 3.3)	\$/acre-ft
ρ_m	Constant	
σ_{jt}	A Heavyside function; 1 designates that capital must be be provided to build a canal to supply imported water to reservoir j in year t; 0 indicates that no capital is to be provided	
ω_{ijt}	Net revenue coefficient for the energy generated by reservoir j in month i of year t	\$/kW-hr
σ_m	Maximum allowable BOD concentration in beginning of link m at saturation oxygen concentration	mg/l
ϕ_m	Slope of curve of maximum allowable BOD concentration versus initial dissolved oxygen deficit for arc m	

Subscripts

i	The month of operation
j	The project (dam/treatment plant) in question
m	The link (river, canal) in question
t	The year of operation

Superscript

T	Transpose of a matrix

REFERENCES

Ackoff, R. L. (1961). "Progress in Operations Research," p. 367. Wiley, New York.

Alexander, T. (1967). A wild plan for South America's wilds. *Fortune* **73,** 148 et seq.

Amir, R. (1967). Optimum Operation of a Multi-reservoir Supply System. Rep. No. EEP-24. Dept. Eng.-Econ. Planning, Stanford Univ., Palo Alto, California.

Balas, E. (1965). An additive algorithm for solving linear programs with zero-one variables. *Oper. Res.* **13,** 517–546.

Barkin, D., and King, T. (1970). "Regional Economic Development: The River Basin Approach in Mexico." Cambridge Univ. Press, London and New York.

Bellman, R. E., and Dreyfus, S. E. (1962). "Applied Dynamic Programming." Princeton Univ. Press, Princeton, New Jersey.

Benders, J. F. (1962). Partitioning procedures for solving mixed-variables programming problems. *Numer. Math.* **4,** 238–252.

Beveridge, G. S., and Schechter, R. S. (1970). "Optimization: Theory and Practice." McGraw-Hill, New York.

Bowen, W. (1965). Water shortage is a frame of mind. *Fortune* **71,** 144 et seq.

Bower, B. T. (1962). A simplified river-basin system for testing methods and techniques of analysis, *in* "Design of Water Resources Systems" (A. B. Maass, ed.), Chapter 7. Harvard Univ. Press, Cambridge, Massachusetts.

Burkley, J. W., *et al.* (1965). On the Water-Resource Problems of Latin America. Tech. Rep. No. 87. Hydrodyn. Lab., Dept. of Civil Eng., M.I.T., Cambridge, Massachusetts.

Butcher, W. S. (1968). Stochastic dynamic programming and the assessment of risk. *Proc. Nat. Symp. Anal. of Water-Resource Syst., Denver, Colorado 1968.*

Butcher, W. S., Haimes, Y. Y., and Hall, W. A. (1969). Dynamic programming for the optimal sequencing of water supply projects. *Water Resour. Res.* **5,** 1196–1207.

Buttermore, P. M. (1966). Water use in the petroleum and natural gas industries. *U.S. Bur. Mines Inform. Cir.* IC **8284**.

California State Water Resour. Board (1957). California Water Plan Bull. 3. State of California, Sacramento.

Camp, T. R. (1963). "Water and Its Impurities." Van Nostrand-Reinhold, Princeton, New Jersey.

Carter, W. A., Schulthess, H., Salazar, E., and Guzman, R. S. (1967). Financing water projects in Latin America. *Int. Conf. Water for Peace, Washington, D. C., 1967,* Paper No. 714.

Caulfield, H. F., Jr. (1967). Techniques of water resources planning, water law and institutions. *Annu. Water Resour. Res. Conf., 2nd, 1967*, p. 29, Office of Water Resour. Res., U.S. Dept. of Interior, Washington, D.C.

Center of Latin American Studies (1962). "Statistical Abstracts of Latin America—1962." Center of Latin American Studies, Univ. of California, Los Angeles.

Ciriacy-Wantrup, S. V. (1955). Benefit-cost analysis and public resources development. *J. Farm Econ.* **3,** 688.

Clawson, M., and Knetsch, J. L. (1966). "Economics of Out-door Recreation." Published for Resources for the Future by Johns Hopkins Press, Baltimore, Maryland.

Clay, C. (1969). A review of the Texas water plan. *In* "Issues and Attitudes in Contemporary Developments in Water Law" (C. W. Johnson and J. Lewis, eds.). Center for Res. in Water Resour., Univ. of Texas Press, Austin.

Clough, D. J., and Bayer, M. B. (1970). Optimal waste treatment and pollution abatement benefits on a closed river system, *in* "Applications of Mathematical Programming Techniques" (E. M. L. Beale, Ed.). English Universities Press, London.

Colombia Inform. Service (1969). Centralized development program to secure $329 millions in external project loans for Colombia in 1969. *Colombia Today* **4,** No. 1–2.

Cord, J. (1964). A method for allocating funds to investment projects when returns are subject to uncertainty. *Management Sci.* **10,** 335–341.

Criddle, W. D. (1958). Methods of computing consumptive use of water. *Proceed. ASCE J. Irrigation Drainage Div.* **84** (**IR-1**), 1–27.

Dantzig, G. B. (1957). Discrete-variable extremum problems. *Oper. Res.* **8,** 275 et seq.

Dantzig, G. B. (1963). "Linear Programming and Extensions." Princeton Univ. Press, Princeton, New Jersey.

Davis, P. S., and Ray, T. L. (1969). A branch-bound algorithm for the capacitated facilities location problem, *Nav. Res. Logistics Quart.* **16,** 331–344.

Day, H. J., Dolbear, Jr., T., and Kamien, M. (1965). Regional water quality management. *Annu. Meeting AWRA 1st, Univ. of Chicago, Chicago, Illinois, 1965.*

Deininger, R. A. (1965a). Water quality management—The planning of economically optimal pollution control systems. Ph.D. Thesis, Northwestern Univ., Evanston, Illinois.

Deininger, R. A. (1965b). Water quality management—The planning of economically optimal pollution control systems. *Annu. Meeting AWRA, 1st, Univ. of Chicago, Chicago, Illinois, 1965.*

de Lucia, R., McBean, E., and Harrington, J. (1974). System optimization for water quality management in the Saint John River. To be published in *Computers Oper. Res.*

de Neufville, R., and Stafford, J. H. (1971). "Systems Analysis for Engineers and Managers," Chapters 8 and 9. McGraw-Hill, New York.

Ditton, R. B. (1969). The identification and critical analysis of selected literature dealing with the recreational aspects of water resources use, planning, and development. Water Resour. Center, Univ. of Illinois, Urbana.

Dobbins, W. E. (1964). BOD and oxygen relationships in streams. *J. Sanit. Eng. Div. Amer. Soc. Civil Eng.* **90** (SA3).

Drobny, N. L. (ed.) (1970). "Dimensions of Water Management." Battelle Memorial Inst., Columbus, Ohio.

Durbin, E. P., and Kroenke, D. M. (1967). The Out-of-Kilter Algorithm: A Primer. Mem., RM-5472 PR. Rand Corp., Santa Monica, California.

Dysart, B. C., and Hines, W. W. (1969). Development and Application of a Rational Water Quality Model. Final Rep. on OWRR Proj. No. A-012-GA. Completed at

Schools of Ind. and Civil Eng. in Cooperation with Water Resour. Center, Georgia Inst. of Technol., Atlanta.

ECLA (Econ. Commission for Latin America) (1961). Los Recursos Hidraulicos y su Aprovechamiento en America Latina, Venezuela, E/CN.12/593. United Nations, New York.

ECLA (Econ. Commission for Latin America) (1963a). Los Recursos Hidraulicos de America Latina—Reseña y Evaluacion de la Labor Realizada por la Cepal, E/CN.12/650. United Nations, New York.

ECLA (Econ. Commission for Latin America) (1963b). Los Recursos Naturales en America Latina, Su Conocimiento Actual e Investigaciones Necesarias en Este Campo—El Agua, E/CN.12/670/Add. 2. United Nations, New York.

ECLA (Econ. Commission for Latin America) (1964). Los Recursos Hidraulicos de America Latina: Bolivia y Colombia, New York. United Nations, New York.

ECLA (Econ. Commission for Latin America) (1967). Latin American Hydroelectric Potential, Economic Survey of Latin America, **12,** No. 1. United Nations, New York.

Efroymson, M. A., and Ray, T. L. (1965). A Branch-Bound Algorithm for Plant Location. Esso Res. and Eng. Co., Linden, New Jersey.

Evenson, D. E., and Moseley, J. C. (1970). Simulation/optimization techniques for multi-basin water resources planning. *Water Resour. Bull.* **6,** 725–736.

Everett, H. (1963). Generalized Lagrange multiplier method for solving problems of optimum allocation of resources. *Oper. Res.* **11,** 399–417.

Fiering, M. B. (1966). Statistical analysis of stream flow data. Ph.D. Thesis, Harvard Univ., Cambridge, Massachusetts.

Fitch, W. N., King, P. H., and Young, G. K. (1970). The optimization of a multi-purpose water resources system. *Water Resour. Bull.* **6,** 498–518.

Ford, L. R., and Fulkerson, D. R. (1962). "Flows in Networks," pp. 162–169. Princeton Univ. Press, Princeton, New Jersey.

Fulkerson, D. R. (1961). An out-of-kilter method for minimal cost flow problems. *SIAM* (Soc. Indust. Appl. Math.) *J. Appl. Math.* **9,** 18–27.

Garfinkel, R. S., and Nemhauser, G. L. (1972). "Integer Programming." Wiley, New York.

Generoso, E. E. (1966). Optimal capital allocation to the expansion of an existing chemical plant. Ph.D. Thesis, S.U.N.Y., Buffalo, New York.

Glover, F., and Zionts, S. (1965). A note on the additive algorithm of Balas. *Oper. Res.* **13,** 546–549.

Gloyna, E. F., D'Arezzo, A. J., and O'Laoghaire, D. T. (1971). Water Quality Management in a River Basin Undergoing Rapid Economic Growth: A Mathematical Model and a Proposed Methodology for its Solution, Appendix R in Instrumentation for Engineering Management of a Multi-Purpose River Basin System (The Trinity River Basin, Texas): An Interim Report. Center for Res. in Water Resour., Univ. of Texas, Austin, Texas.

Golamb, S. W., and Baumert, L. D. (1965). Backtrack programming. *J. Ass. Comput. Mach.* **12,** 516–524.

Gomory, R. E. (1960). All Integer Programming Algorithm. Res. Rep. RC-189. IBM.

Graves, G. W., Hatfield, G. B., and Whinston, A. (1969). Water pollution control using bypass piping. Water Resour. Res. **5,** 15–47.

Graves, G. W., Pingry, D., and Whinston, A. (1972). Water quality control: nonlinear programming algorithm. Rev. Française d'Automatique, Informat. Recherche Opérationnelle **6** (V-2), 49–78.

Guthrie, B. (1958). "Hydro-Electric Engineering Practice," Vol. I. Blackie, Glasgow and London.

Haimes, Y. Y. (1972). Decomposition and multilevel techniques for water quality control. Water Resourc. Res. **8,** 779–784.

Haimes, Y. Y., Foley, J., and Yu, W. (1971). Computational results for water pollution taxation using multilevel approach. *Annu. Water Resour. Conf., 7th, Washington, D. C., 1971.*

Haimes, Y. Y., Kaplan, M. A. and Husar, M. A., Jr. (1972). A multilevel approach to determining optimal taxation for the abatement of water pollution. *Water Resourc. Res.* **8,** 851–860.

Hall, W. A. (1964). Optimum design of a multipurpose reservoir. *J. Hydraulics Div. Amer. Soc. Civil Eng.* **90** (HY4), 141–149.

Hall, W. A., and Buras, N. (1961). The dynamic programming approach to water resources development. *J. Geophys. Res.* **66,** 517–520.

Hall, W. A., and Dracup, J. A. (1970). "Water Resources Systems Engineering," Chapter 2. McGraw-Hill, New York.

Hall, W. A., and Howell, D. T. (1963). The optimization of single purpose reservoir design with the application of dynamic programming to synthetic hydrology samples. *J. Hydro.* **1,** 355–363.

Hall, W. A., and Shephard, R. W. (1967). Optimum operations for planning of a complex water resources system. *Univ. of California Water Resour. Center Contrib.* No. **122.**

Hall, W. A., Butcher, W. S., and Esogbue, A. (1968). Optimization of the operation of a multi-purpose reservoir by dynamic programming. *Water Resour. Res.* **4,** 471–477.

Hargreaves, G. H. (1957). Irrigation requirements based on climatic data. *J. Irrigation Drainage Div. Amer. Soc. Civil Eng.* **82(IR-3),** Paper 1105.

Hays, Jr., A. J., and Gloyna, E. F. (1969). A Least-Cost Analysis for the Houston Ship Channel. Rep. CRWR-50. Center for Res. in Water Resour., Univ. of Texas, Austin.

Hays, Almond J., Jr. and Earnest F. Gloyna, February 1972. Optimal Water Quality Management for the Houston Ship Channel, *J. Sanitary Engr. Div. ASCE,* **98,** (*SA1*), 195–214.

Hillier, F. S., and Lieberman, G. J. (1967). "Introduction to Operations Research," Chaps. 5, 13, 14, 15. Holden-Day, San Francisco, California.

Himmelblau, D. M. (1972). "Applied Nonlinear Programming." McGraw-Hill, New York.

Hinomoto, H. (1970). Linear programming applied to multi-stage capacity expansion of water treatment-distribution system. *Nat. Meeting ORSA, 38th, Detroit, Michigan,* 1970.

Hirshleifer, J., de Haven, J. C., and Milliman, J. W. (1960). "Water Supply: Economics, Technology and Policy." Univ. of Chicago Press, Chicago, Illinois.

Hooke, R., and Jeeves, T. A. (1961). Direct search solution of numerical and statistical problems. *J. Ass. Comput. Mach.* **8,** 212–229.

Howard, G. T., and Nemhauser, G. L. (1968). Optimal capacity expansion. *Nav. Res. Logistics Quart.* **15,** 535–550.

Howe, C. W., and Linaweaver, M. M. (1967). The impact of price on residential water demand and its relation to system design and price structure. *Water Resour. Res.* **3,** 12–32.

Howe, C. W., Russell, C. S., Young, R. A., and Vaughan, W. J. (1971). "Future Water Demands" Resources for the Future, Inc., Washington, D.C.

Hu, T. C. (1970). "Integer Programming and Network Flows." Addison-Wesley, Reading, Massachusetts.

Hufschmidt, M. M. (1962). Analysis by simulation: Examination of response surface, *in* "Design of Water Resources Systems" (A. B. Maass, ed.), Chapter 10. Harvard Univ. Press, Cambridge, Massachusetts.

Hufschmidt, M. M., and Fiering, M. B. (1966). "Simulation Techniques for Design of Water-Resource Systems," Chapter 1. Harvard Univ. Press, Cambridge, Massachusetts.

IBRD (Int. Bank for Reconstruction and Develop.) (1952). "The Basis of a Development Plan for Colombia." John Hopkins Press, Baltimore, Maryland.

IBRD (Int. Bank for Reconstruction and Develop.) (1967). Water and economic development. *Int. Conf. Water for Peace, Washington, D. C., 1967*, Paper No. 714.

James, J. W. S. (1969). Example of the application of a computer to water resources development. *Comput. Water Resour.* pp. 273–306.

Kalter, R. J., Lord, W. B., Allee, D. J., and Castle, E. N. (1969). Criteria for Federal Evaluation of Resource Investments. Water Resour. and Marine Sci. Center, Cornell Univ., Ithaca, New York.

Kane, J. W. (1967). Monthly Reservoir Evaporation Rates for Texas, 1940 through 1965. Rep. 64. Texas Water Develop. Board, Austin, Texas.

Kaplan, S. (1966). Solutions of the Lorie-Savage and similar integer programming problems by the generalized Lagrange multipliers. *Oper. Res.* **14,** 1130–1136.

Kaufman, A., and Nadler, M. (1966). Water use in the mineral industry. *U.S. Bur. Mines Inform. Cir.* **IC 8285.**

Kerri, K. D. (1965). An economic approach to water quality control. *Annu. Conf. Water Pollut. Contr. Fed., 38th, Atlantic City, New Jersey, 1965.*

Kerri, K. D. (1967). A dynamic model for water quality control. *J. Water Pollut. Contr. Fed.* **39,** 772–786.

King, Jr., J. A. (1967). "Economic Development Projects and their Appraisal." Published for the Int. Bank for Reconstruction and Develop. by Johns Hopkins Press, Baltimore, Maryland. Case Study 1: The Development of Electric Power in the Valle del Cauca (in Colombia).

Kneese, A. V. (1964). "Economics of Regional Water Quality Management." Johns Hopkins Press, Baltimore, Maryland.

Kneese, A. V. (1967). "Approaches to Regional Water Quality Management." Resources for the Future, Inc., Washington, D. C.

Land, A. H., and Doig, A. (1960). An automatic method of solving discrete programming problems. *Econometrica* **28,** 497–520.

Lasdon, L. (1970). "Optimization Theory for Large Scale Systems." Macmillan, New York.

Lawler, E. L., and Wood, D. E. (1966). Branch-and-bound methods: A survey. *Oper. Res.* **14,** 699–719.

Lee, E. S., and Waziruddin, S. (1970). Applying gradient projection and conjugate gradient to the optimum operation of reservoirs. *Water Resour. Bull.* **6,** 713–724.

Lesso, W. G. (1967). Long Range Corporate Planning Model for Major Capital Investments. Tech. Memo. No. 83. Oper. Res. Group, Case Western Reserve Univ., Cleveland, Ohio.

Lesso, W. G. (1969). An extension of the net present value concept to intertemporal investments. *Eng. Economist* **15,** 1 et seq.

Lewis, D. J., and Shoemaker, L. A. (1962). Hydro system power analysis by digital computer. *J. Hydraulics Div. Amer. Soc. Civil Eng.* **88** (HY3), 113–130.

Liebman, J. C. (1969). Water quality management models. Lecture notes from the course "Optimization Techniques in the Planning Design and Operation of Water Resources Systems." UCLA, Los Angeles, California.

Liebman, J. C., and Lynn, W. R. (1966). The optimal allocation of stream dissolved oxygen. *Water Resour. Res.* **2,** 581–591.

Liebman, J. C., and Marks, D. H. (1968). A Balas algorithm for zoned uniform treatment. *J. Sanit. Eng. Div. Amer. Soc. Civil Eng.* **94** (SA4), 585–593.

Little, J. D. C., Murty, K. G., Sweeney, D. W., and Karel, C. (1963). An algorithm for the travelling salesman problem. *Oper. Res.* **11,** 972–989.

Lofting, E. M., and McGaughey, P. M. (1968). Economic evaluation of water; Part IV: An input-output analysis of California water requirements. Univ. of California Water Resour. Center Contrib. No. **116.**

Loucks, D. P. (1969). Stochastic Methods for Analyzing River Basins. Water Resour. and Marine Sci. Center, Cornell Univ., Ithaca, New York.

Loucks, D. P., Revelle, C. S., and Lynn, W. R. (1967). Linear programming models for water pollution control. *Management Sci.* **14,** B-166–B-181.

Lowery, Jr., R. L. (1960). Monthly Reservoir Evaporation Rates for Texas. Bull. 6006. Texas Board of Water Eng., Austin, Texas.

Lynn, W. R., Logan, J. A., and Charnes, A. (1962). System analysis for planning wastewater treatment plants. *J. Water Pollut. Contr. Fed.* **34,** 565–581.

Maass, A. B. (1962). System design and the political process: a general statement, *in* "Design of Water-Resource Systems" (A. B. Maass, *et al.*, eds.) Chapter 15. Harvard Univ. Press, Cambridge, Massachusetts.

Maass, A., and Hufschmidt, M. M. (1962). Methods and techniques of analysis of the multiunit, multipurpose water-resource system, *in* "Design of Water Resources Systems" (A. B. Maass, ed.). Harvard Univ. Press, Cambridge, Massachusetts.

McCarthy, E. A. (1966). "Dictionary of American Politics," p. 154. Penguin Books, Baltimore, Maryland.

McDaniels, L. L. (1960). Consumptive Use of Water by Major Crops in Texas. Bull. 6019. Texas Board of Water Eng., Austin, Texas.

McLaughlin, R. T. (1967). Experience with preliminary system analysis for river basins. *Int. Conf. on Water for Peace, Washington, D. C., 1967,* Paper 408.

McQuade, W. (1970). Global earth-shapers in complex competition. *Fortune* **76,** 78 et seq.

Manne, A. S. (1960). Product Mix Alternatives: Flood Control, Electric Power, and Irrigation. Cowles Foundation Discuss. Paper No. 95, published at Yale Univ., New Haven, Connecticut.

Marglin, S. (1962). Economic factors affecting system design, *in* "Design of Water Resources Systems" (A. B. Maass, ed.). Harvard Univ. Press, Cambridge, Massachusetts.

Massé, P. (1962). "Optimal Investment Decisions: Rules for Action and Criteria for Choice." Prentice-Hall, Englewood Cliffs, New Jersey.

Massé, P., and Gibrat, R. (1957). Applications of linear programming to investments in the electric power industry. *Management Sci.* **3.**

Meier, G. M. (1964). "Leading Issues in Development Economics," p. 202. Oxford Univ. Press, London and New York.

Merewitz, L. (1966). Recreational benefits of water resource development. *Water Resour. Res.* **2,** 625–640.

Milligan, J. H. (1970). Optimizing Conjunctive Use of Groundwater and Surface Water. PRWG-42-4T. Utah Water Res. Lab., Utah State Univ., Logan, Utah.

Mobasheri, F. (1968). Economic evaluation of a water resources development project in a developing economy. Univ. of California Water Resour. Center Contrib. No. **126**.

Mobasheri, F., and Harboe, R. C. (1970). A two-stage optimization model for design of a multipurpose reservoir. *Water Resourc. Res.* **6,** 22–31.

Morin, T. (1970). Optimal scheduling of water resource projects. *Nat. Meeting ORSA, 38th, Detroit, Michigan,* 1970.

Morin, T. L. (1973). Optimal sequencing of capacity expansion projects. *J. Hydraulics Div. Amer. Soc. Civil Engr.* **99** (HY9) 1605–1622.

Morrice, H. A. W., and Allan, W. N. (1959). Planning for the ultimate development of the Nile Valley. *Proc. Inst. Civil Eng.* **14,** 101–156.

Moseley, J. C., Puentes, C. C., and Weiss, A. O. (1969). Simulation/optimization and the Texas water plan. *Hydraulic Tech. Group, Amer. Soc. Civil Eng., Lubbock, Texas, 1969.*

Nat. Ass. of Mfr. and the Chamber of Commerce of the U.S. (1965). Water in Industry.

Nat. Eng. Co. Surface Water Quality in Texas. Rep. prepared for the Texas Water Develop. Board, Austin.

Nayak, S. C., and Arora, S. R. (1970). A separable programming approach to the determination of optimal capacities in a multi-reservoir system. *Nat. Meeting ORSA, 38th, Detroit, Michigan, 1970.*

Nemhauser, G. L., and Ullmann, Z. (1969). Discrete dynamic programming and capital allocation. *Management Sci.* **15,** 494–505.

Nunamaker, J. F., Pomeranz, J., and Whinston, A. B. (1974). A data base planning system for problems of world concern. To be published in *Computers Oper. Res.*

O'Laoghaire, D. T. (1974). Water quality management in a river basin undergoing rapid economic growth. To be published in *Computers Oper. Res.*

O'Laoghaire, D. T., and Himmelblau, D. M. (1971). Optimal capital investment in the expansion of an existing water resources system. *Water Resour. Bull.* **7,** 1194–1209.

O'Laoghaire, D. T., and Himmelblau, D. M. (1972). Modeling and sensitivity analysis for planning decisions in water resources expansion. *Water Resourc. Bull.* **8,** 653–668.

Olson, S. H. (1966). Some conceptual problems of interpreting the value of water in humid regions. *Water Resour. Res.* **2,** 1–11.

Orlob, G. T. (1970). Planning the Texas water system. *Annu. Water Resour. Res. Conf., 5th, Washington, D.C., 1970,* pp. 171–186.

Pan Amer. Health Organization/World Health Organization (1963). Environmental Sanitation in the National Plan of Social and Economic Development. Ten Year Plan for Water Supply in Urban and Rural Areas of Latin America, Washington, D. C.

Parsons, R. M. (Co.) (1964). North Amer. Water and Power Alliance (NAWAPA) Brochure 606-2934-16. R. M. Parsons Co., Los Angeles, California.

Paviani, D. A. (1969). A new method for the solution of the general non-linear programming problem. Ph.D. Thesis, Univ. of Texas, Austin.

Penman, H. L. (1956). Estimating evaporation. *Trans. Amer. Geophys. Union* **37,** 43–50.

Peterson, C. C. (1967). Computational experience with variants of the Balas algorithm applied to the selection of R and D projects. *Management Sci.* **13,** 735–750.

Posada F., Antonio J., and Anderson de Posada, J. (1966). "La CVC: Un Reto al Subdesarrollo y Tradicionalismo." Ediciones Tercer Mundo, Bogotá, Colombia.

Revelle, C., Loucks, D., and Lynn, W. R. (1967). A management model for water quality control. *J. Water Poll. Control Fed.* **39** (6).

Revelle, C. S., Loucks, D. P., and Lynn, W. R. (1968). Linear programming applied to water quality management. *Water Resour. Res.* **4,** 1–9.

Rogers, P. (1969). A game theory approach to the problems of international river basins. *Water Resour. Res.* **5**, 749–760.

Schwerg, A., and Cole, J. A. (1968). Optimal control of linked reservoirs. *Water Resour. Res.* **4**, 479–497.

Sen, S. (1969). "United Nations in Economic Development: Need for a New Strategy." Oceana Publ., Dobbs Ferry, New York.

Sewell, W. R., and Bower, B. T. (eds.) (1968). "Forecasting the Demands for Water." Queens Printer, Ottawa, Canada.

Sirles, III, J. E. (1968). Application of Marginal Economic Analysis to Reservoir Recreation Planning. Res. Rep. No. 12. Univ. of Kentucky Water Resour. Inst., Lexington, Kentucky.

Sobel, M. J. (1965). Water quality improvement programming problems. *Water Resour. Res.* **1**, 477–487.

State of California Dept. of Water Resour. (1964). Water Use by Manufacturing Industries in California 1957–1959. Bull. No. 124. Sacramento State Printing Office, Sacramento, California.

Steiner, P. O. (1959). Choosing among alternative public investments in the water resources field. *Amer. Econ. Rev.* **69**, 893–916.

Stewart, H., and Fraser, R. H. (1969). *In* "Reservoirs: Problems and Conflicts" (E. E. N. Castle, ed.). Water Resour. Res. Inst., Oregon State Univ., Corvallis, Oregon.

Swanson, H. S. (1970). A network flow analysis of water allocation decisions in a river system and their effects on estuarine ecology. Ph.D. Thesis, Univ. of Texas, Austin.

Texas Water Develop. Board (1968). The Texas Water Plan. Texas Water Develop. Board, Austin.

Texas Water Develop. Board (1969). A Completion Report on System Simulation for Management of a Total Water Resource. Texas Water Develop. Board, Austin.

Texas Water Develop. Board (1970). Biochemical Oxygen Demand, Dissolved Oxygen, Selected Nutrients, and Pesticide Records of Texas Surface Waters, 1968. Rep. 108. Texas Water Develop. Board, Austin.

Texas Water Develop. Board (1971). Stochastic Optimization and Simulation Techniques for Management of Regional Water Resources Systems. Rep. 131. Texas Water Develop. Board, Austin.

Thomann, R. V., and Marks, D. H. (1967). Results from a systems analysis approach to the optimum control of estuarine water quality. *Adv. Water Poll. Res.* **3**, 29–42.

Thomann, R. V., and Sobel, M. J. (1964). Estuarine water quality management and forecasting. *J. Sanit. Eng. Div. Amer. Soc. Civil Eng.* **90** (**SA5**), 90–97.

Thomas, H. A., and Revelle, R. (1966). On the efficient use of the high Aswan Dam for hydropower and irrigation. *Management Sci.* **12**, B296–B311.

Tinbergen, J. (1964). "Central Planning." Yale Univ. Press, New Haven, Connecticut.

Tussey, R. C. (1967). Analysis of Reservoir Recreation Benefits. Publ. No. 2. Univ. of Kentucky Water Resour. Inst., Lexington, Kentucky.

United Nations (1958). Water for Industrial Use. Dept. of Econ. and Soc. Affairs, United Nations, New York.

United Nations (1962). "Estudios Sobre La Electricidad en America Latina," Vol. I. United Nations, New York.

United Nations (1964). "Estudios Sobre La Electricidad en America Latina," Vol. II. United Nations, New York.

United Nations (1970). "Food and Agricultural Yearbook." United Nations, Food and Agricultural Organization, Rome, Italy.

U.S. Census of Agriculture (1969). Vols. I and III. Bur. of Census, U.S. Dept. of Commerce, Washington, D.C.

U.S. Census of Manufacturers (1972). Published by U.S. Dept. of Commerce, Bur. of Census.

U.S. Congr., Senate (1962). Policies, Standards, and Procedures in the Formulation, Evaluation, and Review of Plans for Use and Development of Water and Related Land Resources. Senate Document No. 97. 2nd Session, 87th Congr.

U.S. Corps of Eng. (1962). Delaware River Basin, New York, New Jersey, Pennsylvania and Delaware. House Document 522. 87th Congress.

U.S. Dept. of Health, Educ., and Welfare, Public Health Serv. (1963). Statement, Water Pollution Control and Quality Management Programs of the Pacific Northwest, Data Attachments, Vol. 2. U.S. Dept. of Health, Educ., and Welfare, Public Health Serv., Portland, Oregon.

U.S. Geol. Surv. (1957). Compilation of Surface Water Records Ohio River Basin, 1950. Water Supply Paper 1305. Dept. of the Interior. Washington, D.C.

U.S. Geol. Surv. (1961). Surface Water Supply of the United States; Part 5, Hudson Bay and Upper Mississippi River Basins, 1959. Water Supply Paper 1928. Dept. of the Interior. Washington, D.C.

U.S. Geol. Surv. Circ. (1951). *U.S. Geol. Surv. Circ.* No. **115**.

U.S. Geol. Surv. Circ. (1956). *U.S. Geol. Surv. Circ.* No. **398**.

U.S. Geol. Surv. Circ. (1961). *U.S. Geol. Surv. Circ.* No. **456**.

U.S. Govt. (1958). Proposed Practices for Economic Evaluation of River Basin Projects. Rep. to the Inter-Agency Committee on Water Resources, (Green Book). U.S. Govt. Printing Office, Washington, D.C.

U.S. Study Commission—Texas (1961). Inventory of Present Water Quality in the Study Area. U.S. Public Health Serv. Washington, D.C.

Wallace, J. R. (1966). Linear Programming Analysis of River-Basin Development. Sc.D. Thesis, M.I.T., Cambridge, Massachusetts.

Water Resour. Center (1968). Rep. 126. Water Resour. Center, Univ. of California, Berkeley.

Water Resour. Council (1969). Nation's Water Resources. Washington, D.C.

Weaver, R. M. (1963). Preliminary Design of Concrete Dams by Means of Computers. Portland Cement Ass., Chicago, Illinois.

Weingartner, H. M. (1966). Capital budgeting of interrelated projects: Survey and synthesis. *Management Sci.* **12,** 485–516.

Weiss, A. O., and Beard, L. R. (1970). A multi-basin planning strategy. *Amer. Water Resour. Conf., 6th, Las Vegas, Nevada, 1970.*

White, G. F. (1965). "Strategies of American Water Development." Univ. of Michigan Press, Ann Arbor.

White, G., *et al.* (1969). "Water and Choice in the Colorado River Basin." Nat. Acad. of Sci. Washington, D.C.

Wilde, D. J. (1964). "Optimum Seeking Methods." Prentice-Hall, Englewood Cliffs, New Jersey.

Wilde, D. J. (1965). Strategies for optimizing macrosystems. *Chem. Eng. Progr.* **61,** 86–93.

Wilde, D. J., and Beightler, C. S. (1967). "Foundations of Optimization." Prentice-Hall, Englewood Cliffs, New Jersey.

Woolsey, R. E. D. (1969). An Application of Integer Programming to Optimal Water Resource Allocation for the Delaware River Basin. Ph.D. Thesis, Univ. of Texas Austin.

Young, G. K. (1967). Finding reservoir operating rules. *J. Hydraulics Div. Amer. Soc. Civil Eng.* **93**(HY6), 297–321.

Young, G. K., and Pisano, M. A. (1970). Nonlinear programming applied to regional water resources planning. *Water Resour. Res.* **6,** 32–42.

Young, G. K., Moseley, J. C., and Evenson, D. E. (1969). Time sequencing of element construction in a multi-reservoir system. *Joint Annu. AWRA, 5th, and Annu. Water for Texas Conf., 14th, San Antonio, Texas, 1969.*

Young, G. K., Moseley, J. C., and Evenson, D. E. (1970). The sequencing of element construction in a multi-reservoir system. *Water Resour. Bull.* **6,** 528–541.

SUPPLEMENTARY READING

Chapter 1

Buras, N. (1965). A three-dimensional optimization problem in water-resources engineering, *Operational Res. Quarterly* **16,** 419–427.

Buras, N. (1966). Dynamic programming in water resources development. *Advan. Hydrosci.* **3,** 376–412.

Buras, N. (1972). "Scientific Allocation of Water Resources." Amer. Elsevier, New York.

Clyde, C. G., *et al.* "Applications of Operations Research Techniques for Allocation of Water Resources in Utah." Utah State Univ., Logan, Utah.

de Lucia, R. J. (1971). "Systems Analysis in Water Resources Planning." Meta Systems, Inc., Cambridge, Massachusetts. (PB-204-374).

Dorfman, R. (1962). Mathematical models: The multistructure approach, *in* "Design of Water Resources Systems" (A. B. Maass, ed.), Chapter 13. Harvard Univ. Press, Cambridge, Massachusetts.

Drobny, N. L. (1968). Water resources systems analysis—An overview. *Proc. Amer. Water Resour. Conf., 4th, 1968,* pp. 534–558. Amer. Water Resour. Ass., Chicago, Illinois.

Eckman, D. P. (ed.) (1961). "Systems: Research and Design." Wiley, New York.

Eckstein, O. (1958). "Water-Resource Development." Harvard Univ. Press, Cambridge, Massachusetts.

Ellis, D. O., and Ludwig, F. J. (1962). "Systems Philosophy." Prentice-Hall, Englewood Cliffs, New Jersey.

Fiering, M. B. (1967). "Streamflow Synthesis." Harvard Univ. Press, Cambridge, Massachusetts.

Greenberg, H. (1971). "Integer Programming." Academic Press, New York.

Hufschmidt, M. M., and Fiering, M. B. (1966). "Simulation Techniques for Design of Water Resources Systems." Harvard Univ. Press, Cambridge, Massachusetts.

Kerr, J. A. (1972). Multireservoir Analysis Techniques in Water Quantity Studies. *Water Resourc. Bull.* **8,** 871–880. (A valuable source of reference for the Russian literature.)

Lasdon, L. S. (1970). "Optimization Theory for Large Systems." Macmillan, New York.

Maass, A. (ed.) (1962). "Design of Water Resources Systems." Harvard Univ. Press, Cambridge, Massachusetts.

McKean, R. N. (1958). "Efficiency in Government Through Systems Analysis." Wiley, New York.

Marglin, S. A. (1963). "Approaches to Dynamic Investment Planning." North-Holland Publ., Amsterdam.

Mesarovic, M. D. (1964). "Views on General Systems Theory." Wiley, New York.

Schaeffer, K. H. (1962). "The Logic of An Approach to the Analysis of Complex Systems." Project No. IMU-3546. Stanford Res. Inst., Stanford, California.

Shamir, U. (1972). Optimizing the operation of Israel's water supply. *Technol. Rev.* **74**, 41–48.

Thompson, R. G. (1971). "Forcasting Water Demands." Nat. Water Commission, Washington, D. C. (PB-206-491).

Water Resour. Council (1964). "Policies, Standards, and Procedures in the Formulation Evaluation and Review of Plans for Use and Development of Water and Related Land Resource," Suppl. No. 1, "Evaluation Standards for Primary Outdoor Recreation Benefits." Water Resour. Council, Washington, D. C.

Water Resour. Res. Council, Washington, D. C.:

December 1971. Proposed Principles and Standards for Planning Water and Related Land Resources. (PB-209-187).

September 1971. Revision of Completed Regional or River Basin Plans. (PB-209-147).

February 1971. Water and Related Land Resources Management. The Challenge Ahead. (PB-209-156).

June 1971. *Proc. Annu. Conf. State and Fed. Water Officials, 5th, Des Moines, Iowa, June, 1971.* (PB-209-157).

July 1970. Water and Related Land Resources Planning. (PB-209-145).

July 1970. Summary: Federal Agency Technical Comments on the Special Task Force Report Entitled 'Procedures for Evaluation of Water and Related Land Resource Projects.' (PB-209-172).

July 1970. Summary and Index. Public Response to the Special Task Force Report Entitled Procedures for Evaluation of Water and Related Land Resource Projects'. (PB-209-173).

July 1970. Findings and Recommendations. (PB-209-174).

July 1970. Principles for Planning Water and Land Resources. (PB-209-175).

July 1970. Standards for Planning Water and Land Resources. (PB-209-176).

July 1970. A Summary Analysis of Nineteen Tests of Proposed Evaluation Procedures on Selected Water and Land Resource Projects. (PB-209-177).

Wismer, D. A., ed. (1971). "Optimization Theory for Large—Scale Systems." McGraw-Hill, New York.

Chapter 2

Buras, N., and Hall, W. A. (1961). An analysis of reservoir capacity requirements for conjunctive use of surface and groundwater storage, *in* "Symposium of Athens," pub. 57, pp. 556–563. Int. Assoc. Sci. Hydrology.

Cannel, G. H. (1962). Irrigation efficiency as it influences water requirements of crops, *in* "Water Requirements of Crops," pp. 47–56. Amer. Soc. Agr. Engr., St. Joseph, Michigan.

Corps of Eng. U.S. Army (1960). Routing of floods through river channels. "Engineering and Design Manual."

Haimes, Y. Y. (1971). Modeling and control of the pollution of water resources systems via the multilevel approach. *Water Resour. Bull.* **7,** 93–101.

Haimes, Y. Y., Foley, J., and Yu, W. (1972). Computational results for water pollution taxation using multilevel approach. *Water Resour. Bull.* **8,** 761–772.

Hall, W. A., and Roefs, T. G. (1966). Hydropower project output optimization. *J. Power Div. Amer. Soc. Civil Eng.* **92**(P01), 67–79.

Hufschmidt, M. M., and Fiering, M. B. (1966). "Simulation Techniques for Design of Water Resources Systems." Harvard Univ. Press, Cambridge, Massachusetts.

Kuiper, E. (1965). "Water Resources Development." Butterworth, London.

Lewis, D. J., and Shoemaker, L. A. (1962). Hydro system power analysis by digital computer. *J. Hydraulics Div. Amer. Soc. Civil Eng.* **88,** (HY3), 113–130.

Linsley, R. K., and Franzini, J. B. (1964). "Water Resources Engineering." McGraw-Hill, New York.

Romm J. (1969). The value of reservoir recreation. Tech. Rep. 19. Water Resour. Marine Sci. Center, Cornell Univ., Ithaca, New York.

Chapter 3

Agin, N. (1966). Optimum seeking with branch and bound. *Management Sci.* **13,** B176–B185.

Balas, E. (1968). A note on the branch and bound principle. *Oper. Res.* **16,** 442–445.

Berretta, J. C., and Mobasheri, F. (1972). An optimal strategy for capacity expansion. *Engineering Economist,* **17,** 79–98.

Burkov, V. N., and Lovetskii, S. E. (1968), Methods for the solution of extremal problems of combinatorial type. *Automat. Remote Contr.* (*USSR*) **29,** 1785–1806.

Erlenkotter, D. (1973). Sequencing of Interdependent Hydroelectric Projects, *Water Resourc. Res.* **9,** 21–27.

Ford, L. R., and Fulkerson, D. R. (1962). "Flows in Networks." Princeton Univ. Press, Princeton, New Jersey.

Frank, H., and Frisch, I. T. (1971). "Communication, Transmission and Transportation Networks." Addison-Wesley, Reading, Massachusetts. (A unified rigorous treatment of deterministic and probabilistic networks.)

Fulkerson, D. R. (1961). An out-of-kilter method for minimal cost flow problems. *SIAM* (*Soc. Ind. Appl. Math.*) *J. Appl. Math.* **9,** 19–27.

Geoffrion, A. M. (1970). Elements of large-scale mathematical programming. *Management Sci.* **16,** 652–691.

Texas Water Develop. Board (1971). Stochastic Optimization and Simulation Techniques for Water Management of Regional Water Resources Systems. Rep. No. 131. Texas Water Develop. Board, Austin.

Thompson, III, I. G. (1970). Modifications of the out-of-kilter algorithm to handle networks with leak from one or more nodes: A limited case. M.S. Thesis, Univ. of Texas, Austin.

Chapter 5

Beard, L. R. (1965), Use of interrelated records to simulate stream flow. *J. Hydraulics Div. Amer. Soc. Civil Eng.* **91** (HY5), 13.

Charnes, A., and Cooper, W. W. (1959). Chance constrained programming. *Management Sci.* **6,** 73–80.

Chen, M. J. K., Erickson, L. E., and Fan, L. T. (1970). Considerations of sensitivity and parameter uncertainty in optimal process design. *Ind. Eng. Chem. Process Des. Develop.* **9,** 514–521.

Chow, V. T., ed. (1964). Statistical and probability analysis of hydrological data, *in* "Handbook of Applied Hydrology." McGraw-Hill, New York.

Hall, W. A., and Dracup, J. A. (1970). "Water Resources Engineering." McGraw-Hill, New York.

Loucks, D. P. (1969). Stochastic Methods for Analyzing River Basin Systems. Tech. Rep. No. 16. Water Resour. and Marine Sci. Center, Cornell Univ., Ithaca, New York.

Meier, W. L., Weiss, A. O., Puentes, C. C., and Moseley, J. C. (1970). Sensitivity analysis, a necessity in water planning. *Annu. Amer. Water Resour. Ass. Conf., 6th, Las Vegas, Nevada, 1970.*

Tomovic, R. (1963). "Sensitivity Analysis of Dynamic Systems." McGraw-Hill, New York.

Wilkie, D. F., and Perkins, W. R. (1969). Essential parameters in sensitivity analysis. *Automatica J. IFAC* **5,** 191–197.

Chapter 6

Baxter, S. S. (1965). Economic consideration of water pollution control. *J. Water Pollut. Contr. Fed.* **37,** 1363–1369.

Davis, R. K. (1968). "The Range of Choice in Water Management." Johns Hopkins Press, Baltimore, Maryland.

Dept. of Sci. and Ind. Res., U.K. (1964). Effects of Polluting Discharges Upon the Thames Estuary. Water Pollut. Res. Tech. Paper No. 11. HM Stationery Office, London.

Dysart, B. C., and Hines, W. W. (1969). Development and Application of a Rational Water Quality Planning Model. Georgia Inst. of Technol., Atlanta. (PB-184-835).

Fox, I. K., and Herfindahl, O. C. (1964). "Attainment of Efficiency in Satisfying Demands for Water Resources." Resources for the Future, Inc., Washington, D.C.

Frankel, R. J., and Hansen, W. W. (1968). Biological and physical response in a fresh water dissolved oxygen model, *in* "Advances in Water Quality Improvement" (E. F. Gloyna and W. W. Eckenfelder, eds.), pp. 126–140. Univ. of Texas Press, Austin, Texas.

Gysi, M., and Loucks, D. P. (1969). A Selected Annotated Bibliography on the Analysis of Water Resource Systems. Publ. No. 25. Water Resour. and Marine Sci. Center, Cornell Univ., Ithaca, New York.

Hamilton, H. R. (1968). Some economic aspects of water-quality enhancement. *Meeting AIChE, St. Louis, Missouri, 1968.*

Jaworski, N., Weber, W. J., and Deininger, R. A. (1968). Optimal Release Sequences for Water Quality Control in Multiple-Purpose Reservoir Systems. Dept. of Civil Eng., Univ. of Michigan, Ann Arbor.

Kneese, A. V. (1964). "The Economics of Regional Water Quality Management." Johns Hopkins Press, Baltimore, Maryland.

Kneese, A. V., and Bower, B. T. (1968). "Managing Water Quality: Economics, Technology, Institutions." Johns Hopkins Press, Baltimore, Maryland.

Liebman, J. C., and Lynn, W. R. (1966). The optimal allocation of stream dissolved oxygen. *Water Resour. Res.* **2,** 581–591.

O'Connor, D. J., St. John, J. P., and DiToro, D. M. (1968). Water quality analysis of the Delaware River estuary. *J. Sanit. Eng. Div. Amer. Soc. Civil Eng.* **94**(SA6), 1225.

Pisano, W. C. (1968). River Basin Simulation Program. U. S. Dept. of the Interior, Fed. Water Pollut. Contr. Administration, Washington, D.C.

Shih, C. S., Miller, D. S., and Meier, W. L. (1970). A systems approach to municipal water quality management. *Nat. Meeting Oper. Res. Soc. Amer., 37th, Washington, D.C., 1970.*

Streeter, H. W., and Phelps, E. B. (1925). A Study of the Pollution and Natural Purification of the Ohio River. Bull. 146. U. S. Public Health Service, Washington, D.C.

U. S. Dept. of Health, Educ., and Welfare, Public Health Service (1965). *Symp. Streamflow Regulation for Quality Control, Cincinnati, Ohio, 1965.*

AUTHOR INDEX

Numbers in italics refer to the pages on which the complete references are listed.

SUBJECT INDEX

A 4
B 5
C 6
D 7
E 8
F 9
G 0
H 1
I 2
J 3